C++语言程序设计教程

(第4版)习题解答与实验指导

◎ 沈显君 杨进才 编著

清华大学出版社
北京

内 容 简 介

本书是《C++语言程序设计教程(Visual C++ 2010 版)》(第 4 版)的配套教学用书。全书包括 4 章内容：第 1 章包括《C++语言程序设计教程(Visual C++ 2010 版)》(第 4 版)中的全部习题及其详细解答,题型涵盖程序设计语言考试的主要题型(填空题、选择题、程序填空题、程序分析题、编程题);第 2 章介绍目前较为流行的 C++语言三大开发环境——Visual C++、CodeBlocks 和 Linux C++的程序开发过程,包括编辑、编译、调试过程;第 3 章精心设计了 11 个上机实验题,并给出参考答案,供读者循序渐进地学习与上机练习;第 4 章以一个"自助图书借阅系统"的开发为例,详细描述了面向对象的分析、面向对象的设计、面向对象的编程的面向对象的软件开发过程。

本书可以单独使用,适合作为高等学校 C++语言的辅助教材和全国计算机等级考试的参考书。

本书封面贴有清华大学出版社防伪标签,无标签者不得销售。
版权所有,侵权必究。举报: 010-62782989, beiqinquan@tup.tsinghua.edu.cn。

图书在版编目(CIP)数据

C++语言程序设计教程(第 4 版)习题解答与实验指导/沈显君,杨进才编著.—北京: 清华大学出版社,2022.1(2025.1重印)
21 世纪高等学校计算机类专业核心课程系列教材
ISBN 978-7-302-57305-0

Ⅰ. ①C… Ⅱ. ①沈… ②杨… Ⅲ. ①C++语言-程序设计-高等学校-教材 Ⅳ. ①TP312.8

中国版本图书馆 CIP 数据核字(2021)第 005887 号

策划编辑: 魏江江
责任编辑: 王冰飞
封面设计: 刘　键
责任校对: 徐俊伟
责任印制: 沈　露

出版发行: 清华大学出版社
　　　网　　址: https://www.tup.com.cn, https://www.wqxuetang.com
　　　地　　址: 北京清华大学学研大厦 A 座　　邮　编: 100084
　　　社 总 机: 010-83470000　　　　　　　　　邮　购: 010-62786544
　　　投稿与读者服务: 010-62776969, c-service@tup.tsinghua.edu.cn
　　　质量反馈: 010-62772015, zhiliang@tup.tsinghua.edu.cn
　　　课件下载: https://www.tup.com.cn,010-83470236
印 装 者: 三河市少明印务有限公司
经　　销: 全国新华书店
开　　本: 185mm×260mm　　印　张: 17　　字　数: 434 千字
版　　次: 2022 年 1 月第 1 版　　　　　　　　印　次: 2025 年 1 月第 4 次印刷
印　　数: 3901～4900
定　　价: 39.80 元

产品编号: 089400-01

前　言

党的二十大报告中指出：教育、科技、人才是全面建设社会主义现代化国家的基础性、战略性支撑。必须坚持科技是第一生产力、人才是第一资源、创新是第一动力，深入实施科教兴国战略、人才强国战略、创新驱动发展战略，这三大战略共同服务于创新型国家的建设。高等教育与经济社会发展紧密相连，对促进就业创业、助力经济社会发展、增进人民福祉具有重要意义。

本书是《C++语言程序设计教程（Visual C++ 2010 版）》（第 4 版）的配套教学用书。

学习计算机编程语言不仅要掌握语言本身的语法规则，更重要的是正确运用语言进行编程。本书通过各种题型的题目加强读者对概念、语法的掌握，通过对编程环境的介绍以及实验辅导教会读者如何编程。

本书共分四章。

第 1 章 习题解答，不仅给出习题的答案，而且对习题进行了详细的讲解。本章采用填空题、选择题、简答题、改错题、编程题等多种题型，供学习者理解巩固知识点使用，编程题供上机练习使用。题目与教材的知识点紧密结合，涵盖了《C++语言程序设计教程（Visual C++ 2010 版）》（第 4 版）的所有知识点，其中，习题 1～习题 12 涵盖了等级考试所有知识点。题目中，填空题的难度稍微简单。其他题目有易有难，总体难度适中，既没有太简单的题目，也没有太难的题目。编程题中大多数是 C++语言程序设计课程的经典题目，参考《C++语言程序设计教程（Visual C++ 2010 版）》以及其他课本，读者是可以独立完成的。本章给出了完整的题目，因此本书也可以单独使用。

第 2 章 C++开发环境使用指南，介绍了 Windows 系统环境下的 Visual C++ 2010、CodeBlocks 两大主流 C++编译器的使用，包括集成环境下的环境设置、编辑、编译、调试、运行、查帮助各个环节的介绍。

Linux 是一个使用越来越多的操作系统。随着软件保护的力度的加大，Linux 将会成为主流的教学操作系统，Linux 下的 C++将会成为主要的 C++编程环境。因此，本书介绍了 Linux 下 C++编辑器 Emacs、编译器 g++、调试器 GDB 的使用。

C++ 11 新标准在原 C++标准的基础上增加了大量实用的新特性，本版的主教材中对 C++ 11 新标准中的常用元素进行了介绍，本书介绍的几个开发环境对 C++ 11 新标准均提供了支持。

第 3 章 C++上机实验指导。与其他编程语言一样，掌握 C++编程的秘诀是"上机（编程）、上机、再上机"。本章配合课本的内容，精心设计了 11 个实验，基本上每个实验对应教材的一章。每个实验给出了实验目的、实验内容、实验题目。实验题目选取的是经典的趣味性强的题目,提高学生的编程兴趣。对实验题目给出了参考解答,还就题目的内容提出改进的思考。通过不断地编程训练,提高学生的编程能力。

第 4 章 C++综合应用实验,以一个"自助图书借阅系统"的开发为例,详细描述了面向对象的分析、面向对象的设计、面向对象的编程的面向对象的软件开发过程。

本书可以单独使用,适合作为高等学校 C++语言的辅助教材和全国计算机等级考试的参考书。

本书第 1 章部分内容以及第 3 章由沈显君教授编写;第 1 章部分内容以及第 2 章、第 4 章由杨进才教授编写;第 1 章部分内容由胡珀博士编写。

王敬华副教授对全书的风格、内容提出了建设性的建议,对格式的编排等细节方面也提出了宝贵的意见。在此表示由衷的感谢。

由于本书作者水平有限,书中难免有错误、疏漏和不妥之处,恳请读者批评指正。

编 者

2021 年 8 月于武昌南湖

目 录

第1章 习题解答 ·· 1

    1.1 习题1解答 ·· 1

    1.2 习题2解答 ·· 7

    1.3 习题3解答 ·· 28

    1.4 习题4解答 ·· 48

    1.5 习题5解答 ·· 52

    1.6 习题6解答 ·· 82

    1.7 习题7解答 ·· 111

    1.8 习题8解答 ·· 137

    1.9 习题9解答 ·· 156

    1.10 习题10解答 ·· 167

    1.11 习题11解答 ·· 179

    1.12 习题12解答 ·· 183

第2章 C++开发环境使用指南 ·· 190

    2.1 Visual C++使用指南 ·· 190

        2.1.1 启动 Visual C++ ·· 190

        2.1.2 创建工程 ·· 190

        2.1.3 编辑源程序 ·· 192

        2.1.4 程序的编译与运行 ·· 194

        2.1.5 程序的调试 ·· 196

        2.1.6 多文档工程 ·· 197

    2.2 CodeBlocks 使用指南 ··· 198

        2.2.1 CodeBlocks 的安装与配置 ··· 198

        2.2.2 编辑源程序 ·· 200

        2.2.3 程序的编译与运行 ·· 204

        2.2.4 查帮助 ··· 205

        2.2.5 程序的调试 ·· 206

        2.2.6 多文档工程 ·· 210

2.3 Linux GNU g++ 上机编程指南 210
  2.3.1 使用 EMACS 编辑源程序 210
  2.3.2 g++编译器的使用 215
  2.3.3 程序的运行 216
  2.3.4 查帮助 216
  2.3.5 GDB 调试器的使用 216

第 3 章 C++上机实验指导 219

3.1 上机实验题 219
  3.1.1 实验 1 上机环境和 C++基础编程练习 219
  3.1.2 实验 2 控制结构编程练习 219
  3.1.3 实验 3 函数编程练习 220
  3.1.4 实验 4 构造数据类型编程练习 220
  3.1.5 实验 5 类与对象编程练习 220
  3.1.6 实验 6 继承与派生编程练习 221
  3.1.7 实验 7 多态性编程练习 221
  3.1.8 实验 8 类模板编程练习 221
  3.1.9 实验 9 输入/输出流与文件系统编程练习 221
  3.1.10 实验 10 string 类字符串处理编程练习 222
  3.1.11 实验 11 异常处理编程练习 222
3.2 上机实验题参考解答 222
  3.2.1 实验 1 上机环境和 C++基础编程练习 222
  3.2.2 实验 2 控制结构编程练习 223
  3.2.3 实验 3 函数编程练习 224
  3.2.4 实验 4 构造数据类型编程练习 226
  3.2.5 实验 5 类与对象编程练习 231
  3.2.6 实验 6 继承与派生编程练习 233
  3.2.7 实验 7 多态性编程练习 235
  3.2.8 实验 8 类模板编程练习 238
  3.2.9 实验 9 输入/输出流与文件系统编程练习 240
  3.2.10 实验 10 string 类字符串处理编程练习 242
  3.2.11 实验 11 异常处理编程练习 243

第 4 章 C++综合应用实验 245

4.1 系统分析 245
4.2 系统设计 246
4.3 系统实现 248
本章小结 262

第1章

习题解答

1.1 习题1解答

1. 填空题

(1) 面向对象的方法将现实世界中的客观事物描述成具有属性和行为的**对象**,抽象出共同属性和行为,形成**类**。

(2) C++程序开发通常要经过5个阶段,包括**编辑**、**编译**、**连接**、**运行**、**调试**。首先是**编辑**阶段,任务是**编辑源程序**,C++源程序文件通常带有**.cpp**扩展名。接着使用**编译器**对源程序进行**编译**,将源程序翻译为机器语言代码(目标代码),过程分为词法分析、语法分析、代码生成3个步骤。

在此之前,**预编译器**会自动执行源程序中的**预处理指令**,将其他源程序文件包括到要编译的文件中,以及执行各种文字替换等。

连接器的功能就是将目标码同缺失函数的代码连接起来,将这个"漏洞"补上,生成**可执行文件**。程序运行时,可执行文件由操作系统装入内存,然后CPU从内存中取出程序执行。若程序运行过程中出现了错误,还需要对程序进行**调试**。

(3) 对象与对象之间通过**消息**进行相互通信。

(4) **类**是对具有相同属性和行为的一组对象的抽象;任何一个对象都是某个类的一个实例。

(5) **多态性**是指在一般类中定义的属性或行为被特殊类继承之后可以具有不同的数据类型或表现出不同的行为。

(6) 面向对象的软件开发过程主要包括**面向对象的方法分析**、**面向对象的设计**、**面向对象的编程**、**面向对象的测试**和**面向对象的维护**。

(7) C++提供**名字空间(namespace)**将相同的名字放在不同空间中来防止命名冲突。

(8) ♯ include < iostream >是一条**预处理**指令(语句),在**编译(或预处理)**时由**编译器(或预编译器)**执行,其功能是**将 iostream 文件包含(复制)到指令处**。

(9) C++中使用 **cin** 作为标准输入流对象,通常代表键盘,与提取操作符 **>>** 连用;使用 **cout** 作为标准输出流对象,通常代表显示设备,与 **<<** 连用。

2. 简答题

(1) 简述机器语言、汇编语言、高级语言的特点。

【答】 机器语言是计算机直接执行的语言,由二进制的0和1构成的一系列指令组成;汇编语言是机器语言的助记符;高级语言是接近人的自然语言习惯的编程语言,通过编译变成机器语言。

(2) 结构化语言与面向对象的语言是截然分开的吗？

【答】 不是截然分开的，面向对象的程序设计中也包含过程，含有结构化的思想。

(3) C++语言是纯粹的面向对象的程序设计语言吗？

【答】 不是。C++语言是在 C 语言的基础上引入面向对象程序设计思想形成的，它保留了某些面向过程的程序设计特征。

(4) C 语言编写的程序不加修改就可以在 C++编译器中编译吗？

【答】 绝大多数 C 语言程序都可以在 C++编译器中编译，因为 C++语言兼容 C 语言。

(5) C++的源程序是什么类型的文件？如何在 Word 中进行编辑？

【答】 C++源程序是文书文件（文本文件）；在 Word 中编辑，存为扩展名为.cpp 的文本文件。在 Windows 的写字板、记事本中都可以编辑。

(6) 如何将一个 C++源程序变成可执行程序？产生的各类文件的扩展名是什么？

【答】 通过编译变成带扩展名.obj 的目标文件；再通过连接变成带扩展名.exe 的可执行文件。

(7) 如果要求不使用 include 包含头文件，有什么办法使程序正常编译运行？

【答】 在相应的文件夹（子目录）中找到需包含的头文件，将头文件复制到包含处。

(8) 下列程序中如有错误与不妥当之处请指出。

```
/// *************************************************
*       程序文件名：p1_2.cpp                         *
************************************************* /
Using namespace std
# include < iostream >;
using std::endl;
int main()
    float num1, num2, num3;                //定义 3 个数
    cin << num1 << num2 << num3;
    cout >> "The average is:" >> setw(30) >>(num1 + num2 + num3)/3 >> endl;
    return 0;
}
```

【答】 错误处标号如下：

```
①/// *************************************************
*       程序文件名：p1_2.cpp                         *
************************************************* /
②Using namespace std③
# include < iostream >;④
using std::endl;⑤
int main() ⑥
    float num1, num2, num3;                //定义 3 个数
    cin ⑦ << num1 << num2 << num3;
    cout >> ⑧ "The average is:" ⑨ >> setw(30) ⑩>>(num1 + num2 + num3)/3 >> endl;
    return 0;
}
```

① 行注释符号//将块注释的头/＊注释掉了，使得块注释的尾＊/没有相应的头与之匹配；
② 关键字 Using 中包含了大写字母，应改为 using；
③ using namespace std 不是预处理指令，要以分号结尾；
④ ♯include ＜iostream＞是预处理指令，不能以分号结尾，而且要作为程序的开头行；
⑤ 使用了 using namespace std，就不必单独使用 std∷endl；
⑥ int main()后少了{；
⑦ cin 应与提取操作符＞＞连用；
⑧ cout 应与插入操作符＜＜连用；
⑨ "The average is："为全角引号，应改为英文；
⑩ 使用 setw(30)应包含头文件 iomanip。

修改后的正确程序如下：

```
/*************************************************
 *      程序文件名：p1_2.cpp                      *
 *************************************************/
# include <iostream>
# include <iomanip>
using namespace std;
int main(){
    float num1,num2, num3;                //定义3个数
    cin >> num1 >> num2 >> num3;
    cout <<"The average is:"<< setw(30) <<(num1 + num2 + num3)/3 << endl;
    return 0;
}
```

3. 选择题

(1) C++语言属于(　　)。

  A. 机器语言　　　　B. 低级语言　　　　C. 中级语言　　　　D. 高级语言

【答】　D

(2) C++语言程序能够在不同操作系统下编译、运行，说明 C++具有良好的(　　)。

  A. 适应性　　　　B. 移植性　　　　C. 兼容性　　　　D. 操作性

【答】　B

【注解】　对于计算机硬件，一般使用"兼容"一词，对于程序使用"移植"。

(3) ♯include 语句(　　)。

  A. 总是在程序运行时最先执行　　　　B. 按照在程序中的位置顺序执行

  C. 在最后执行　　　　D. 在程序运行前就执行了

【答】　D

【注解】　♯include 是预处理指令，在编译时就执行了，没有对应的机器指令。

(4) C++程序运行时，总是起始于(　　)。

  A. 程序中的第一条语句　　　　B. 预处理命令后的第一条语句

  C. main()　　　　D. 预处理指令

【答】　C

(5) 下列说法正确的是(   )。

    A. 用 C++语言书写程序时,不区分大小写字母

    B. 用 C++语言书写程序时,每行必须有行号

    C. 用 C++语言书写程序时,一行只能写一个语句

    D. 用 C++语言书写程序时,一个语句可分几行写

【答】 D

(6) 在下面概念中,不属于面向对象编程方法的是(   )。

    A. 对象　　　　　　B. 继承　　　　　　C. 类　　　　　　D. 过程调用

【答】 D

(7) 下列程序的运行结果为(   )。

```cpp
#include <iostream>
#include <iomanip>
using namespace std;
int main()
{
    cout << setprecision(4)
         << setw(3)
         << hex
         << 100/3.0
         <<", ";
    cout << 24 << endl;
    return 0;
}
```

    A. 3.333e+001,18　　B. 33.33,18　　C. 21,18　　D. 33.3,24

【答】 B

4. 程序填空题

为了使下列程序能顺利运行,请在空白处填上相应的内容:

```cpp
#include_____①
#include_____②
_____③_____;
_____④_____;
_____⑤_____;
int main()
{
    float i, j;
    cin_____⑥_____i_____⑥_____j;
    cout_____⑦_____setw(10)_____⑦_____i * j;
    return 0;
}
```

【答】 程序如下：

```cpp
#include <iostream>
#include <iomanip>
using std::cin;
using std::cout;
using std::setw;
int main()
{
    float i, j;
    cin >> i >> j;
    cout << setw(10) << i * j;
    return 0;
}
```

5. 编程题

（1）编写一个程序，输出用 * 组成的菱形图案。

【答】 程序如下：

```cpp
#include <iostream>
#include <iomanip>
using namespace std;
int main()
{
    cout << setw(16) <<" * "<< endl;
    cout << setw(17) <<" *** "<< endl;
    cout << setw(18) <<" ***** "<< endl;
    cout << setw(19) <<" ******* "<< endl;
    cout << setw(20) <<" ********* "<< endl;
    cout << setw(19) <<" ******* "<< endl;
    cout << setw(18) <<" ***** "<< endl;
    cout << setw(17) <<" *** "<< endl;
    cout << setw(16) <<" * "<< endl;
}
```

或

```cpp
#include <iostream>
#include <iomanip>
using namespace std;
int main()
{
    cout <<"     *    "<< endl;
    cout <<"    ***   "<< endl;
    cout <<"   *****  "<< endl;
    cout <<"  ******* "<< endl;
    cout <<" ********* "<< endl;
    cout <<"  ******* "<< endl;
```

```
    cout <<"     *****   "<< endl;
    cout <<"      ***    "<< endl;
    cout <<"       *     "<< endl;
    return 0;
}
```

(2) 编写一个程序,输入任意十进制数,将其以八进制、十六进制的形式输出。

【答】 程序如下:

```
#include <iostream>
#include <iomanip>
using namespace std;
int main()
{
    int i;
    cout <<"输入十进制数:";
    cin >> i;
    cout <<"八进制:"<< oct << i <<"十六进制:"<< hex << i << endl;
    return 0;
}
```

(3) 仿照本章例题设计一个程序,输入两个数,将它们相除,观察为无限循环小数时按精度从小到大输出的结果。

【答】 程序如下:

```
#include <iostream>
#include <iomanip>
using namespace std;
int main()
{
    double i,j;
    cout <<"输入两个数:";
    cin >> i >> j;
    cout << setprecision(1);
    cout <<"precision(1):"<< i <<"/"<< j <<" = "<< i/j << endl;
    cout << setprecision(2);
    cout <<"precision(2):"<< i <<"/"<< j <<" = "<< i/j << endl;
    cout << setprecision(3);
    cout <<"precision(3):"<< i <<"/"<< j <<" = "<< i/j << endl;
    cout << setprecision(8);
    cout <<"precision(8):"<< i <<"/"<< j <<" = "<< i/j << endl;
    cout << setprecision(9);
    cout <<"precision(9):"<< i <<"/"<< j <<" = "<< i/j << endl;
    cout << setprecision(10);
    cout <<"precision(10):"<< i <<"/"<< j <<" = "<< i/j << endl;
    return 0;
}
```

运行结果：

```
输入两个数:2000  3↙
precision(1):2e+003/3 = 7e+002
precision(2):2e+003/3 = 6.7e+002
precision(3):2e+003/3 = 667
precision(8):2000/3 = 666.66667
precision(9):2000/3 = 666.666667
precision(10):2000/3 = 666.6666667
```

1.2 习题 2 解答

1. 填空题

(1) C++的基本数据类型可分为五大类，即<u>逻辑型（或布尔型）</u>、<u>字符型</u>、<u>整型</u>、<u>实型</u>、<u>空值型</u>，分别用关键字 **bool**、**char**、**int**、**float/double**、**void** 定义，长度分别为 <u>1</u>、<u>1</u>、<u>4</u>、<u>4/8</u>、<u>不定</u>字节。整型、字符型的默认符号修饰为 **signed**。

【注解】另一种填法为：

C++的基本数据类型可分为五大类，即<u>逻辑型（或布尔型）</u>、<u>字符型</u>、<u>整型</u>、<u>浮点型</u>、<u>双精度型</u>，分别用关键字 **bool**、**char**、**int**、**float**、**double** 定义，长度分别为 <u>1</u>、<u>1</u>、<u>4</u>、<u>4</u>、<u>8</u> 字节。整型、字符型的默认符号修饰为 **signed**。

(2) 十进制数值、八进制数值、十六进制数值的前缀分别为 <u>1～9</u>、<u>0</u>、<u>0x（或 0X）</u>。

(3) 在 C++预定义的常用转义序列中，在输出流中用于换行、空格的转义序列分别为<u>\n</u>、<u>\t</u>。

(4) 布尔型数值只有两个，即 **true**、**false**。在 C++的算术运算式中，它们分别当作 <u>1</u>、<u>0</u>。

(5) 字符由<u>单引号</u>''括起来，字符串由<u>双引号</u>" "括起来。字符只能有 <u>1</u> 个字符，字符串可以有<u>多</u>个字符。空串的表示方法为"<u>\0</u>"（或<u>" "</u>）。

(6) 关系运算符操作数的类型可以是<u>任何基本数据类型</u>，对其中的<u>实数</u>类型数不能进行直接比较。

(7) &&与||表达式<u>按从左到右</u>的顺序进行计算，以 && 连接的表达式，如果左边的计算结果为 **false（或 0）**，右边的计算不需要进行，就能得到整个逻辑表达式的结果：**false**；以 || 连接的表达式，如果左边的计算结果为 **true（或非 0）**，就能得到整个逻辑表达式的结果：**true**。

(8) >>运算符将一个数右移 n 位，相当于将该数<u>除以</u> 2^n，<<运算符将一个数左移 n 位，相当于将该数<u>乘以</u> 2^n。

(9) 所有含赋值运算的运算符的左边要求是<u>左值</u>。

(10) 前置++、--的优先级<u>高</u>于后置++、--。

(11) 按操作数数目分，运算符的优先级从高到低排列为<u>单目</u>、<u>双目</u>、<u>三目</u>，按运算符的性质分，优先级从高到低排列为<u>算术</u>、<u>移位</u>、<u>关系</u>、<u>按位</u>、<u>逻辑</u>。

(12) 在表达式中，会产生副作用的运算符有<u>++</u>、<u>--</u>、<u>各类赋值</u>。

(13) 函数执行过程中通过 **return** 语句将函数值返回，当一个函数不需要返回值时需要使用 **void** 作为类型名。

(14) 在 C++程序中，如果函数定义在后，调用在先，需要进行<u>函数原型声明</u>，告诉编译器函数的<u>（返回）类型</u>、<u>函数名</u>、<u>形式参数</u>。其格式和定义函数时的函数头的形式基本相同，它必须以**分号**结尾。

(15) 函数参数传递过程的实质是将实参值通过**堆栈**——传送给形参。

(16) 递归程序分两个阶段执行,即**调用**、**回代**。

(17) 带 inline 关键字定义的函数为**内联函数**,在**编译**时将函数体展开到所有调用处。内联函数的好处是节省**执行时间**开销。

(18) 函数名相同,但对应形参表不同的一组函数称为**重载函数**,参数表不同是指**参数个数**、**类型不同**。

(19) 确定对重载函数中函数进行绑定的优先次序为**精确匹配**、**对实参的类型向高类型转换后的匹配**、**实参类型向低类型及相容类型转换后的匹配**。

(20) 内联函数的展开、重载函数的绑定、类模板的实例化与绑定均在**编译**阶段进行。

2. 选择题

(1) 下列选项中,均为常用合理的数值的选项是(　　)。

　　A. ·25　　1L　　0Xfffe　　　　　　B. '好!'　　3333333333　　−01U
　　C. 10^8　　'\''　　'\x'　　　　　　D. 08　　FALSE　　1e+08

【答】　A

(2) 上述选项中,均为不合理(能通过编译,但不提倡使用)的数值的选项是(　　)。

【答】　B

【注解】　'好!'中用单引号引起来的不止一个字符;3333333333 超过整型数的范围;−01U 从符号看是一个负数,从后缀看是一个无符号数,相互矛盾。

(3) 上述选项中,均为不合法(不能通过编译)的数值的选项是(　　)。

【答】　C

【注解】　10^8 不能在编辑器中输入;'\''、'\x'为不合法的转义序列。08 中的 8 不是八进制的合法数字;FALSE 不是 bool 型常量,false 为 bool 型常量。1e+08 是合法的常量。

(4) 下列选项中,均为合法的标识符的选项是(　　)。

　　A. program　　a&b　　2me　　　　B. ccnu@mail　　C++　　a_b
　　C. π　　变量a　　a b　　　　　　D. ____Line　　_123　　Cout

【答】　D

【注解】　A 中的 a&b,B 中的 ccnu@mail、C++ 含有不能构成标识符的符号 &、@、+;2me 以数字开头,它们均为不合法的标识符。

(5) 上述选项中,均为不合法的标识符的选项是(　　)。

【答】　C

【注解】　π 是一个数学符号;变量a 以汉字开头,以汉字开头与含有汉字的都不能作为标识符;a 和 b 中间有空格,标识符中不能含空格。

(6) 若定义"short int i=32769;","cout << i;"的输出结果为(　　)。

　　A. 32 769　　　B. 32 767　　　C. −32 767　　　D. 不确定的数

【答】　C

【注解】　32 769 为整型常量,十六进制形式为 0x00008001,赋给短整型后,截取低 16 位 0x8001,对应的十进制值为 −32 767。

(7) 若定义"char c='\78';",则变量 c(　　)。

　　A. 包含1个字符　　　　　　　　　B. 包含两个字符

　　　　C. 包含 3 个字符　　　　　　　　D. 定义不合法

【答】　D

【注解】　'\78'中的 8 不是八进制的合法数字,'\78'不是一个合法的八进制转义序列。

(8) 若定义"int a=7; float x=2.5, y=4.7;",则 x+a%3 * static_cast<int>(x+y)%2/4 的值为(　　)。

　　　　A. 2.5　　　　　　B. 2.75　　　　　　C. 3.5　　　　　　D. 0.0

【答】　A

【注解】　依次计算"a%3=1; static_cast<int>(x+y)=7; 7%2/4=0; x+0=2.5。"

(9) 设 i 为 int 型、f 为 float 型,则 10 + i + 'f'的数据类型为(　　)。

　　　　A. int　　　　　　B. float　　　　　　C. double　　　　　　D. char

【答】　A

【注解】　'f'为 char 类型,10、i 为 int 型。

(10) 设变量 f 为 float 型,将 f 小数点后第 3 位四舍五入,保留小数点后两位的表达式为(　　)。

　　　　A. (f * 100+0.5)/100
　　　　B. (f * 100+0.5)/100.0
　　　　C. (int)(f * 100+0.5)/100.0
　　　　D. (int)(f * 100+0.5)/100

【答】　C

【注解】　D 的结果为整数,A、B 的结果保留了小数点后若干位。

(11) 下列运算要求操作数必须为整型的是(　　)。

　　　　A. /　　　　　　B. ++　　　　　　C. !=　　　　　　D. %

【答】　D

(12) 若变量已正确定义并具有初值,下列表达式合法的是(　　)。

　　　　A. a:=b++　　　B. a=b+3=c++　　　C. a=b+++=c　　　D. a=b++, b=c

【答】　D

【注解】　A 中的 := 不是 C++中的运算符；B、C 中的 b+3、b++不是左值。

(13) 6 种基本数据类型的长度排列正确的是(　　)。

　　　　A. bool = char < int ≤ long = float < double
　　　　B. char < bool = int ≤ long = float < double
　　　　C. bool < char < int < long < float < double
　　　　D. bool < char < int < long = float < double

【答】　A

【注解】　bool、char、int、long、float、double 的长度分别为 1、1、4(2)、4、4、8 字节。

(14) 若变量 a 是 int 类型,执行"a='A'+1.6;",正确的叙述为(　　)。

　　　　A. a 的值是字符 C
　　　　B. a 的值是浮点型
　　　　C. 不允许字符型数与浮点型数相加
　　　　D. a 的值是'A'的 ASCII 码值加上 1

【答】　D

【注解】　a 的值由 double 型转换成 int 型。

(15) 判断 char 型变量 c 是否为英文字母的表达式为(　　)。

　　　　A. 'a'<=c<='z' && 'A'<=c<='Z'
　　　　B. 'a'<=c&&c<='z' || 'A'<=c&&c<='Z'

C. 'a'<=c<='z' || 'A'<=c<='Z'

D. ('a'<=c||c<='z') && ('A'<=c||c<='Z')

【答】 B

【注解】 A、C 中'a'<=c<='z'的运算顺序为('a'<=c)<='z'。

(16) 下列表达式中没有副作用的是( )。

  A. cout<<i++<<i++   B. a=(b=1)+=2

  C. a=(b=1)+2   D. c=a*b+ ++b

【答】 C

【注解】 A 中的 i++<<i++ 因计算顺序不同有不同的结果；B 中的 a=(b=1)+=2 等价于 a=(b=1)=(b=1)+2；D 中的 c=a*b+ ++b 等价于 c=a*b+(b=b+1)。

(17) 下列语句中的 x 和 y 都是 int 型变量,其中错误的语句是( )。

  A. x=y++;   B. x=++y;

  C. (x+y)++;   D. ++x=y;

【答】 C

(18) 对于 if 语句中的表达式的类型,下面描述正确的是( )。

  A. 必须是关系表达式

  B. 必须是关系表达式或逻辑表达式

  C. 必须是关系表达式或算术表达式

  D. 可以是任意表达式

【答】 D

【注解】 可以是任意表达式甚至是常量。

(19) 以下错误的 if 语句是( )。

  A. if(x>y) x++;

  B. if(x == y) x++;

  C. if(x<y) {x++; y--;}

  D. if(x!= y) cout<<x else cout<<y;

【答】 D

【注解】 cout<<x 不是一个完整的语句,应加上分号。

(20) 以下程序的输出结果为( )。

```
int main() {
    int a(20),b(30),c(40);
    if(a>b)a=b, b=c, c=a;
    cout<<"a = "<<a<<",b = "<<b<<",c = "<<c;
    return 0;
}
```

  A. a=20,b=30,c=40   B. a=20,b=40,c=20

  C. a=30,b=40,c=20   D. a=30,b=40,c=30

【答】 A

(21) 以下程序的输出结果为( )。

```
int main() {
```

```
        int a(1),b(3),c(5),d(4),x(0);
        if(a < b)
            if(c < d) x = 1;
            else if(a < c)
                if (b < d) x = 2;
                else x = 3;
            else x = 6;
        else x = 7;
            cout << x;
        return 0;
    }
```

    A. 1   B. 2   C. 3   D. 6

【答】 B

(22) 以下程序的输出结果为(　　)。

```
    int main() {
        int x(1),a(0),b(0);
        switch(x) {
        case 0: b++;
        case 1: a++;
        case 2: a++; b++;
        }
        cout <<"a = "<< a <<",b = "<< b;
        return 0;
    }
```

    A. a=2, b=1   B. a=1, b=1   C. a=1, b=0   D. a=2, b=2

【答】 A

(23) 以下程序的输出结果为(　　)。

```
    int main()
    {
        int a(15),b(21),m(0);
        switch (a % 3)
        {
            case 0: m++;break;
            case 1: m++;
                switch(b % 2)
                {
                    default: m++;
                    case 0: m++; break;
                }
        }
        cout << m;
        return 0;
    }
```

    A. 1   B. 2   C. 3   D. 4

【答】 A

(24) 与"y=(x>0 ? 1 : x<0 ? -1 : 0);"的功能相同的 if 语句是（   ）。

  A. if(x>0) y=1;
   else if (x<0) y=-1;
   else y=0;

  B. if(x)
   if(x>0) y=1;
   else if (x<0) y=-1;

  C. y = -1;
   if(x)
   if(x>0) y=1;
   else if (x==0) y=0;
   else y=-1;

  D. y = 0;
   if(x>=0)
   if (x>0) y=1;
   else y=-1;

【答】 A

(25) 以下循环的执行次数为（   ）。

  for(int i = 2; i!=0;) cout << i--;

  A. 死循环 B. 0 次 C. 1 次 D. 两次

【答】 D

(26) 下列程序的运行结果是（   ）。

```
int main() {
    int a(1),b(10);
    do {
        b -= a; a++; } while (b-- < 0);
    cout <<"a = "<< a <<",b = "<< b;
    return 0;
}
```

  A. a=3, b=11 B. a=2, b=8 C. a=1, b=-1 D. a=4, b=9

【答】 B

(27) 下面程序段循环执行了（   ）。

  int k = 10;
  while (k = 3) k = k - 1;

  A. 死循环 B. 0 次 C. 3 次 D. 7 次

【答】 A

(28) 语句 while(!E)中的表达式!E 等价于（   ）。

  A. E == 0 B. E!=1 C. E!=0 D. E == 1

【答】 A

(29) 以下程序段（   ）。

  x = -1;
  do {x = x * x;} while (!x);

  A. 是死循环   B. 循环执行 1 次
  C. 循环执行两次   D. 语法错

【答】 B

(30) 与"for(表达式 1;表达式 2;表达式 3)循环体;"功能相同的语句为(   )。

A. 表达式 1;
while(表达式 2){
    循环体;
    表达式 3; }

B. 表达式 1;
while(表达式 2){
    表达式 3;
    循环体; }

C. 表达式 1;
do {
    循环体;
    表达式 3; } while(表达式 2);

D. do {
    表达式 1;
    循环体;
    表达式 3; } while(表达式 2)

【答】 A

(31) 以下循环体的执行次数为(   )。

for(int x = 0,y = 0;(y = 123) && (x < 4); x++);

A. 死循环　　　　B. 次数不定　　　　C. 4 次　　　　D. 3 次

【答】 C

(32) 以下不是死循环的语句为(   )。

A. for(y=0，x=1;x>++y ; x=i++) i=x;
B. for(; ; x++=i);
C. while(1) {x++;}
D. for(i=0; ; i--) sum+=i;

【答】 A

(33) 下列程序的运行结果是(   )。

```
int main() {
    int a(1),b(1);
    for (; a<=100; a++) {
        if (b>=10) break;
        if (b%3==1) {b+=3; continue; }
    }
    cout << a;
    return 0;
}
```

A. 101　　　　B. 6　　　　C. 5　　　　D. 4

【答】 D

(34) 以下循环体的执行次数为(   )。

```
int i(0);
while (i<10) {
    if(i<1) continue;
    if(i==5) break;
    i++;
}
```

A. 死循环　　　　B. 1 次　　　　C. 10 次　　　　D. 6 次

【答】 A

(35) 执行语句序列

```
int n;
cin >> n;
switch(n)
{
    case 1:
    case 2:cout <<'1';
    case 3:
    case 4:cout <<'2';break;
    default:cout <<'3';
}
```

时,若键盘输入1,则屏幕显示（　　）。

　　A. 1　　　　　　　B. 2　　　　　　　C. 3　　　　　　　D. 12

【答】 D

(36) 在下列定义中,正确的函数定义形式为（　　）。

　　A. void fun(void)　　　　　　B. double fun(int x; int y)
　　C. int fun();　　　　　　　　D. double fun(int x,y)

【答】 A

【注解】 B中形参用分号分隔,C中定义以分号结尾,D中y没有类型。

(37) 函数 int fun(int x, int y)的声明形式不正确的为（　　）。

　　A. int fun(int , int);　　　　　B. int fun(int y, int x);
　　C. int fun(int x, int y)　　　　D. int fun(int i, int j);

【答】 C

【注解】 函数声明要以";"结尾。

(38) 在 C++语言中,函数返回值的类型由（　　）。

　　A. return 语句中的表达式类型决定　　B. 调用该函数时的主调函数类型决定
　　C. 调用该函数时系统临时决定　　　　D. 定义该函数时所指定的数据类型决定

【答】 D

(39) 若有函数调用语句：

　　　　fun(a+b, (x, y), (x, y, z));

此调用语句中的实参个数为（　　）。

　　A. 3　　　　　　　B. 4　　　　　　　C. 5　　　　　　　D. 6

【答】 A

【注解】 （x, y）为一个参数。

(40) 在 C++中,关于默认形参值,正确的描述是（　　）。

　　A. 设置默认形参值时,形参名不能省略
　　B. 只能在函数定义时设置默认形参值
　　C. 应该先从右边的形参开始向左边依次设置
　　D. 应该全部设置

【答】 C

(41) 对重载函数的要求,正确的为(　　)。
　　A. 要求参数的个数不同　　　　　　B. 要求参数中至少有一个类型不同
　　C. 要求参数个数相同时类型不同　　D. 要求函数的返回类型不同
【答】　C

(42) 系统在调用重载函数时根据一些条件确定调用哪个重载函数,在下列条件中,不能作为依据的是(　　)。
　　A. 实参个数　　　B. 实参类型　　　C. 函数名称　　　D. 函数类型
【答】　D

(43) 在下列函数原型声明中,错误的是(　　)。
　　A. void Fun(int x＝0,int y＝0);
　　B. void Fun(int x,int y);
　　C. void Fun(int x,int y＝0);
　　D. void Fun(int x＝0,int y);
【答】　D

(44) 若同时定义了以下函数,则 fun(8,3.1)调用的是下列(　　)函数。
　　A. fun (float,int)　　　　　　　B. fun (double,int)
　　C. fun (char,float)　　　　　　D. fun (double,double)
【答】　D

【注解】　函数重载不允许函数名和形参表都相同,仅返回类型不同;绑定的优先次序是精确匹配→对实参的类型向高类型转换后的匹配→实参类型向低类型及相容类型转换后的匹配。本题没有精确匹配,优先进行向高类型转换后的匹配。

3. 简答题

(1) 若定义"int x＝3,y;",那么下列语句执行后 x 和 y 的值分别是多少?
① y＝x＋＋ －1;
② y＝＋＋x－1;
③ y＝x－－ ＋1;

【答】　① "y＝x＋＋－1;"等价于"y＝(x＝x＋1)－1;",执行完后 x、y 的值分别为 4、2。
② "y＝＋＋x－1;"等价于"y＝(x＝x＋1)－1;",执行完后 x、y 的值分别为 4、3。
③ "y＝x－－ ＋1;"等价于"y＝(x＝x－1)＋1;",执行完后 x、y 的值分别为 2、4。

(2) 运行下面的程序段,观察其输出,如将式中的 && 改为 ||,运行结果是什么?

```
int a = 1,b = 2,m = 2,n = 123;
cout <<((m = a>b)&&++n)<< endl;
```

【答】　a＞b 的值为 0,m＝0 的值为 0,短路求值,＋＋n 不计算,因此 n 的值为 123,输出结果为 0。若改成

```
cout <<((m = a>b)||++n)<< endl;
```

a＞b 的值为 0,m＝0 的值为 0,＋＋n 的值为 124,0||124 的值为 1,n 的值为 124,输出结果为 1。

(3) 运行下面的程序,观察其输出。

```
int x,a;
```

```
x = (a = 3 * 5, a * 4), a + 5;
cout <<"x = "<< x <<" a = "<< a << endl;
```

【答】 逗号运算符的优先级比＝低，先计算 a＝3＊5，再计算 a＊4，a＊4 作为(a＝3＊5,a＊4)的返回结果。

输出结果为 x＝60,a＝15。

（4）下列表达式在计算时是如何进行类型转换的？

10/static_cast < float >(3) * 3.14 + 'a' + 10L * (5 > 10)

【答】 类型转换图如图 1-1 所示。

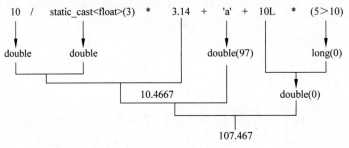

图 1-1 类型转换

4. 程序填空题

（1）下列程序接受从键盘输入的两个数以及＋、－、＊、/运算符，将两个数进行加、减、乘、除，输出运算结果，请填空。

```
int main() {
    char c;
    float a,b,result(0);
    int tag(1);                          //标志,1:合法,0:数据或操作不合法
    cin >> a >> b >> c;
    switch (c) {
        case   ①   : result = a + b; break;
        case   ②   : result = a - b; break;
        case   ③   : result = a * b; break;
        case   ④   :
            if(b == 0) {
                cout <<"divide 0"<< endl;
                tag = 0;
                   ⑤   
            }
            result = a/b;
            break;
           ⑥   
            tag = 0;
            cout <<"invalid operation"<< endl;
               ⑦   
    }
    if(tag)
        cout << result << endl;
    return 0;
}
```

【答】 正确的程序如下：

```cpp
#include <iostream>
using namespace std;
int main() {
    char c;
    float a,b,result(0);
    int tag(1);                           //标志,1:合法,0:数据或操作不合法
    cin >> a >> b >> c;
    switch (c) {
        case    '+'  : result = a + b; break;
        case    '-'  : result = a - b; break;
        case    '*'  : result = a * b; break;
        case    '/'  :
            if(b == 0) {
                cout <<"divide 0"<< endl;
                tag = 0;
                break;
            }
            result = a/b;
            break;
        default:
            tag = 0;
            cout <<"invalid operation"<< endl;
            break;                        //此处可以省略 break
    }
    if(tag)
        cout << result << endl;
    return 0;
}
```

（2）下列程序的功能是输出 100 以内能被 3 整除且个位数为 6 的所有整数，请填空。

```cpp
int main() {
    int i,j;
    for (i = 0;_____①_____;i++) {
        j = i * 10 + 6;
        if (_____②_____) continue;
        cout << j;
    }
    return 0;
}
```

【答】 正确的程序如下：

```cpp
#include <iostream>
using namespace std;
int main() {
    int i,j;
    for (i = 0;i < 10;i++) {
        j = i * 10 + 6;
```

```
        if (j % 3!= 0) continue;
        cout << j;
    }
    return 0;
}
```

(3) 斐波纳契数列有如下特点：第 1、2 个数都是 1，从第 3 个数开始，每个数都是前两个数的和。下列程序的功能是求数列的前 m(m>1) 个数，按每行 5 个数输出。

```
int main() {
    int f1(1),f2(1),m;
    cin >> m;
    cout << f1 <<"\t"<< f2 <<"\t";
    for (int i = 2;i < m; i++) {
        ①_____
        ②_____
        cout << f2 <<"\t";
        if ( ③ ) cout << endl;
    }
    return 0;
}
```

【答】 正确的程序如下：

```
#include <iostream>
using namespace std;
int main() {
    int f1(1),f2(1),m;
    cin >> m;
    cout << f1 <<"\t"<< f2 <<"\t";
    for (int i = 2;i < m; i++) {
        f2 = f1 + f2;
        f1 = f2 - f1;
        cout << f2 <<"\t";
        if ((i + 1) % 5 == 0) cout << endl;
    }
    return 0;
}
```

5. 程序分析题

(1) 写出并分析下列程序的运行结果：

```
#include <iostream>
using namespace std;
void swap(int x,int y) {
    int t;
    t = x,x = y,y = t;
    cout <<"&x:"<< &x <<", &y:"<< &y <<", &t:"<< &t << endl;
}
int main()
{
```

```
    int a = 3, b = 4;
    cout <<"&a:"<< &a <<", &b:"<< &b << endl;
    cout <<"a = "<< a <<", b = "<< b << endl;
    swap(a,b);
    cout <<"a = "<< a <<", b = "<< b << endl;
    return 0;
}
```

【答】 程序结果如下：

```
&a:0013FF7C, &b:0013FF78
a = 3, b = 4
&x:0013FF24, &y:0013FF28, &t:0013FF18
a = 3, b = 4
```

【注解】 函数参数以传值方式不改变实参的值。

（2）下列程序有错误之处，请指出并改正，然后分析改正后的运行结果。

```
#include <iostream>
using namespace std;
int add(int x, int y, int z = 0) {
    cout <<"(int, int, int = 0)\t";
    return x + y; }
int add(int x, char y) {
    cout <<"(int, char)\t";
    return x + y; }
int main(){
    cout << add(9,8)<< endl;
    cout << add(9.0,8.0)<< endl;
    cout << add(9,8.0)<< endl;
    cout << add(9.0,8)<< endl;
    cout << add(9,'A')<< endl;
    cout << add('A','A' - '0')<< endl;
    return 0;
}
```

【答】   cout << add(9.0,8.0)<< endl;
        cout << add(9,8.0)<< endl;
        cout << add(9.0,8)<< endl;   //不能匹配,修改后的程序

```
#include <iostream>
using namespace std;
int add(int x, int y, int z = 0) {
    cout <<"(int, int, int = 0)\t";
    return x + y; }
int add(int x, char y) {
    cout <<"(int, char)\t";
    return x + y; }
double add(double x, double y) {
    cout <<"(double, double)\t";
```

```
        return x + y; }
double add(int x, double y) {
    cout <<"(int, double)\t";
    return x + y; }
double add(double x, int y) {
    cout <<"(double, int)\t";
    return x + y; }
int main(){
    cout << add(9,8)<< endl;
    cout << add(9.0,8.0)<< endl;
    cout << add(9,8.0)<< endl;
    cout << add(9.0,8)<< endl;
    cout << add(9,'A')<< endl;
    cout << add('A','A' - '0')<< endl;
    return 0;
}
```

程序结果如下：

(int, int, int = 0)	17
(double, double)	17
(int, double)	17
(double, int)	17
(int, char)	74
(int, int, int = 0)	82

6. 编程题

（1）摄氏温度与华氏温度的转换公式为：

$$c=\frac{5}{9}(f-32)$$

其中，c 为摄氏温度，f 为华氏温度，写出两者互相转换的表达式，将表达式放到程序中，以整数形式输入一种温度值，以整数形式输出转换后的温度值。

【答】 将华氏温度转换成摄氏温度的程序如下：

```
#include <iostream>
using namespace std;
int main()
{
    int c,f;
    cout <<"输入华氏温度:";
    cin >> f;
    cout <<"摄氏温度为:";
    c = ((5/9.0 * (f - 32)) + 0.5);      //将华氏温度转换成摄氏温度的表达式
    cout << c << endl;
    return 0;
}
```

将摄氏温度转换成华氏温度的程序如下：

```cpp
#include <iostream>
using namespace std;
int main()
{
    int c,f;
    cout <<"输入摄氏温度:";
    cin >> c;
    cout <<"华氏温度为:";
    f = (9/5.0 * c + 32 + 0.5);           //将摄氏温度转换成华氏温度的表达式
    cout << f << endl;
    return 0;
}
```

（2）用三目运算符求 3 个数 x、y、z 的最大者。

【答】　x >= y&&x >= z ? x : y >= x&&y >= z ? y : z

或

x >= y&&x >= z ? x : (y >= x&&y >= z ? y : z)

（3）分别写出引进与不引进第 3 个变量交换两个变量值的表达式（语句）。

【答】　引进第 3 个变量 t 交换变量 x、y 的值：

t = x;
x = y;
y = t;

不引进第 3 个变量交换变量 x、y 的值：

x = x + y;
y = x - y;
x = x - y;

（4）求 n!（n 由键盘输入），当结果将要超出表示范围时退出，显示溢出前的 n 以及结果。

【答】　程序如下：

```cpp
#include <iostream>
using namespace std;
int main() {
    unsigned long n,r(1),next(1);          //r 为结果
    cin >> n;
    for (int i = 1;i <= n; i++)
    {
        r = r * i;
        next = r * (i + 1);
        if(next < r)                       //(n+1)!小于 n!时, 溢出
        {
            cout << i <<"!= "<< r << endl;
            break;
```

```
            }
        }
        if(i > n)                          //不发生溢出
            cout << i - 1 <<"!= "<< r << endl;
        return 0;
}
```

(5) 改写本章将百分制换算成等级分的程序,优化判断。

【答】 一般来说,分数主要分布在 70、80 这一分数段,因此先对分数是否落入这一段进行判断可以减少判断次数。程序如下:

```
#include<iostream>
using namespace std;
int main()
{
    int n;
    cout <<"Enter a score:";
    cin >> n;
    switch(n/10)
    {
        case 7: case 6:
            cout <<"The degree is C"<< endl;
            break;
        case 8:
            cout <<"The degree is B"<< endl;
            break;
        case 9: case 10:
            cout <<"The degree is A"<< endl;
            break;
        default:
            cout <<"The degree is D"<< endl;
    }
    return 0;
}
```

(6) 改写主教材例 3-5 百钱买百鸡程序,减少循环层数以及循环次数,优化循环。

【答】 设 cock(鸡翁)、hen(鸡婆)、chick(鸡雏)各为 x、y、z 只,可以列如下方程:

$$\begin{cases} x+y+z=100 & (1) \\ 5x+3y+\dfrac{z}{3}=100 & (2) \end{cases}$$

利用(1)式可以将三重循环改成二重循环。

```
#include<iostream>
using namespace std;
int main()
```

```
{
    const int cock = 20, hen = 33, chick = 100;         //分别表示鸡翁、鸡婆、鸡雏的最大数
    int j,k;
        for(j = 0; j <= hen; j++)
            for(k = 0; k <= chick; k++)
                if ((5 * (100 - j - k) + 3 * j + k/3) == 100&&k % 3 == 0&&j + k <= 100)
                                        //鸡的个数与钱数必须为整数
                    cout <<"鸡翁、鸡婆、鸡雏各有：\t"<< 100 - j - k <<"\t"<< j <<"\t"<< k << endl;
    return 0;
}
```

(7) 编写一个程序，分别正向、逆向输出 26 个大写英文字母。

【答】 程序如下：

```
#include<iostream>
using namespace std;
int main()
{
    for(int i = 0; i < 26; i++)
        cout << char('A' + i);
    cout << endl;
    for(i = 0; i < 26; i++)
        cout << char('Z' - i);
    return 0;
}
```

(8) 从键盘输入一个整数，判断该数是几位数，逆向输出该数。

【答】 程序如下：

```
#include<iostream>
using namespace std;
int main()
{
    long n,rn(0);                       //rn 为颠倒后的数
    cin >> n;
    for(int i = 0; i < 10; i++)
    {
        if (n > 0)
        {   rn = rn * 10 + n % 10;
            n = n/10;
        }
        else
            break;
    }
    cout << i <<"\t"<< rn << endl;      //i 为未颠倒数的位数
}
```

(9) 从键盘输入一个整数，判断该数是否为回文数。所谓回文数，就是从左向右读与从右向左读都是一样的数，例如 7887、23432 是回文数。

【答】 利用题(7)的程序，将一个数颠倒过来，如果和原来的数相等，即为回文数，否则不是。

程序如下：

```cpp
#include<iostream>
using namespace std;
int main()
{
    int tmp,n,rn(0);                    //rn为颠倒后的数
    cin >> n;
    tmp = n;
    for(int i = 0; i < 10; i++)         //整数最多为10位
    {
        if (tmp > 0)
        {   rn = rn * 10 + tmp % 10;
            tmp = tmp/10;
        }
        else
            break;
    }
    cout << rn << endl;
    if(n == rn)
        cout <<"palindrome number!"<< endl;
    else
        cout <<"not a palindrome number!"<< endl;
    return 0;
}
```

(10) 编写一个程序，按下列公式求圆周率，精确到最后一项的绝对值小于 10^{-8}。

$$\frac{\pi}{4}=1-\frac{1}{3}+\frac{1}{5}-\frac{1}{7}+\cdots$$

【答】 程序如下：

```cpp
#include<iostream>
using namespace std;
int main()
{
    double PI(0);
    for(int i = 1;1.0/(i*2-1)>1e-8;i++)
        PI = PI + 1.0/(i*2-1) * (i%2?1:-1);
    PI = PI * 4;
    cout <<"steps:"<< i <<"\tPI = "<< PI << endl;   //显示计算步骤数与圆周率
    return 0;
}
```

(11) 根据历法，凡是 1、3、5、7、8、10、12 月，每月 31 天；凡是 4、6、9、11 月，每月 30 天；2 月闰年 29 天，平年 28 天。闰年的判断方法如下：

① 如果年号能被 400 整除，此年为闰年；

② 如果年号能被 4 整除，且不能被 100 整除，此年为闰年；

③ 否则不是闰年。

编程输入年、月,输出该月的天数。

【答】 程序如下:

```cpp
#include<iostream>
using namespace std;
int main()
{
    int year,month,day;
    bool leap;
    cout<<"Input year and month:";
    cin>>year>>month;
        if((year%400==0)||(year%4==0&&year%100!=0))
            leap=true;
        else
            leap=false;
        month=(month%13==0?1:month%13);
    switch(month)
    {
      case 1:
      case 3:
      case 5:
      case 7:
      case 8:
      case 10:
      case 12:
          {
            day=31;
            break;
          }
      case 4:
      case 6:
      case 9:
      case 11:
          {
            day=30;
            break;
          }
      case 2:
          if(leap)                          //是否闰月
              day=29;
          else
              day=28;
          break;
    }
    cout<<"days:"<<day<<endl;
    return 0;
}
```

(12) 假定邮寄包裹的计费标准如下表,输入包裹重量以及邮寄距离,计算出邮资。

重量/g	邮资/(元/件)	重量/g	邮资/(元/件)
15	5	60	14(每满1 000km 加收1元)
30	9	≥75	15(每满1 000km 加收1元)
45	12		

* 重量在档次之间按高档计。

【答】 程序如下:

```cpp
#include<iostream>
using namespace std;
int main()
{
    int weight,distance,fee;
    cout<<"Input weight and distance:";
    cin>>weight>>distance;
    if(weight<=15)
        fee=5;
    else if (weight<=30)
        fee=9;
    else if (weight<=45)
        fee=12;
    else if (weight<=60)
        fee=14+distance/1000;
    else
        fee=15+distance/1000;
    cout<<"fee = "<<fee<<endl;
    return 0;
}
```

分析计费标准,发现每 15g 一个档次,也可用 switch…case 语句实现如下:

```cpp
#include<iostream>
using namespace std;
int main()
{
    int weight,distance,fee;
    cout<<"Input weight and distance:";
    cin>>weight>>distance;
    switch ((weight-1)/15) {
    case 0:
        fee=5;
        break;
    case 1:
        fee=9;
        break;
    case 2:
        fee=12;
```

```
                break;
            case 3:
                fee = 14 + distance/1000;
                break;
            default:
                fee = 15 + distance/1000;
        }
        cout <<"fee = "<< fee << endl;
        return 0;
    }
```

(13) 编写一个函数,按不同精度求圆周率。求圆周率的公式如下:

$$\frac{\pi}{4}=1-\frac{1}{3}+\frac{1}{5}-\frac{1}{7}+\cdots$$

【答】 程序如下:

```
#include<iostream>
using namespace std;
double PI(double precision)
{
    double tmp(0);
    for(int i = 1;1.0/(i*2-1)>precision;i++)
        tmp = tmp + 1.0/(i*2-1) * (i%2?1:-1);
    return tmp * 4;
}
int main()
{
    cout << PI(0.00000001)<< endl;
    return 0;
}
```

(14) 设计一个函数,判断一整数是否为素数。

【答】 按素数的定义,用 2~number-1 的所有整数除 number,若能整除,说明 number 不是素数。为了减少循环次数,将除数范围设为 2~sqrt(number)。

```
#include<iostream>
#include<cmath>
using namespace std;
bool IsPrimeNumber(int number)
{
    if(number <= 1)
        return false;
    for(int i = 2;i < sqrt(number);i++)
        if(number % i == 0)
            return false;
    return true;
}
int main()
{
```

```
    int x;
    cout <<"input a number:";
    cin >> x;
    if(IsPrimeNumber(x))
        cout <<"a prime number!"<< endl;
    else
        cout <<"not a prime number!"<< endl;
    return 0;
}
```

(15) 设计一个递归函数,求 x 的 y 次幂。

【答】 对于 y 为正整数的情况,构造递归公式(函数)如下:

$$\begin{cases} \text{power}(x,0) = 1, & y = 0 \\ \text{power}(x,y) = \text{power}(x,y-1) * x, & \text{其他} \end{cases}$$

程序如下:

```
#include <iostream>
using namespace std;
double power(double x, unsigned y)
{
    if(y == 0)
        return 1;
    else
        return x * power(x, y - 1);
}
int main()
{
    double x;
    unsigned y;
    cout <<"input x,y:";
    cin >> x >> y;
    cout <<"power("<< x <<","<< y <<") = "<< power(x,y)<< endl;
    return 0;
}
```

运行结果:

```
input x,y:3.5 9 ↙
power(3.5,9) = 78815.6
```

1.3 习题 3 解答

1. 选择题

(1) 执行以下语句后的输出结果是(   )。

```
enum weekday {sun, mon, tue, wed = 4, thu, fri, sat};
weekday workday = mon;
```

cout << workday + wed << endl;

  A. 6   B. 5   C. thu   D. 编译错

【答】 B

(2) 在 C++ 中引用数组元素时,其数组下标的数据类型允许是( )。

  A. 整型常量   B. 整型表达式

  C. 非浮点型表达式   D. 任何类型的表达式

【答】 C

【注解】 数组下标不允许是浮点型,整型表达式实际上包括整型常量。

(3) 设有数组定义"char array [] = "China";",则数组 array 所占的空间为( )。

  A. 4 个字节 B. 5 个字节 C. 6 个字节 D. 7 个字节

【答】 C

【注解】 其内容为 'C'、'h'、'i'、'n'、'a'、'\0'。

(4) 若有说明"int a[][3] = {1, 2, 3, 4, 5, 6, 7};",则 a 数组高维的大小是( )。

  A. 2   B. 3   C. 4   D. 无确定值

【答】 B

(5) 以下定义语句不正确的是( )。

  A. double x[5] = {2.0, 4.0, 6.0, 8.0, 10.0};

  B. int y[5] = {0, 1, 3, 5, 7, 9};

  C. char c1[] = {'1', '2', '3', '4', '5'};

  D. char c2[] = {1, 2, 3};

【答】 B

【注解】 数组元素超过维数。

(6) 若二维数组 a 有 m 列,则在 a[i][j] 前的元素个数为( )。

  A. j * m + i B. i * m + j C. i * m + j − 1 D. i * m + j + 1

【答】 B

(7) 以下能对二维数组 a 正确初始化的语句是( )。

  A. int a[2][] = {{1, 0, 1}, {5, 2, 3}};

  B. int a[][3] = {{1, 2, 3}, {4, 5, 6}};

  C. int a[2][4] = {{1, 2, 3}, {4, 5}, {6}};

  D. int a[][3] = {{1, 0, 1}, { }, {1, 1}};

【答】 B

(8) 以下不能对二维数组 a 正确初始化的语句是( )。

  A. int a[2][3] = {0};

  B. int a[][3] = {{1, 2}, {0}};

  C. int a[2][3] = {{1, 2}, {3, 4}, {5, 6}};

  D. int a[][3] = {1, 2, 3, 4, 5, 6};

【答】 C

(9) 定义以下变量和数组:

  int k;

  int a[3][3] = {1,2,3,4,5,6,7,8,9};

则下面语句的输出结果是(　　)。

```
for (k = 0; k < 3; k++)
    cout << a[k][2 - k]<<"\t";
```

A. 3 5 7　　　　B. 3 6 9　　　　C. 1 5 9　　　　D. 1 4 7

【答】 A

(10) 在下列定义中,不正确的是(　　)。

A. char a[10] = "china";　　　　B. char a[10], * p = a; p="china";
C. char * a=0;　　　　　　　　　D. int * p=10;

【答】 D

(11) 以下不能正确对字符串赋初值的语句是(　　)。

A. char str[5] = "good!";　　　　B. char str[] = "good!";
C. char str[8] = "good!";　　　　D. char str[5] = {'g', 'o', 'o', 'd'};

【答】 A

(12) 给出以下定义,则正确的叙述为(　　)。

```
char x[] = "abcdefg";
char y[] = {'a', 'b', 'c', 'd', 'e', 'f', 'g'};
```

A. 数组 x 和数组 y 等价　　　　　　B. 数组 x 和数组 y 的长度相同
C. 数组 x 的长度大于数组 y 的长度　　D. 数组 x 的长度小于数组 y 的长度

【答】 C

(13) 以下程序的输出结果是(　　)。

```
int main ()
{
    char st[20] = "hello\0\t\\";
    cout << strlen(st)<<"\t"<< sizeof(st);
    return 0;
}
```

A. 9　9　　　　B. 5　20　　　　C. 13　20　　　　D. 20　20

【答】 B

(14) 下列程序的输出结构是(　　)。

```
#include <iostream>
using namespace std;
int main ()
{
    char a[] = "Hello,World";
    char * ptr = a;
    while ( * ptr)
    {
        if( * ptr >= 'a'&& * ptr <= 'z')
            cout << char( * ptr + 'A' - 'a');
        else cout << * ptr;
        ptr++;
    }
```

```
        return 0;
    }
```
A. HELLO,WORLD B. Hello,World
C. Hello,world D. hello,world

【答】A

(15) 要禁止修改指针 p 本身,又要禁止修改 p 所指向的数据,这样的指针应定义为( )。

A. const char *p="ABCD";
B. char const *p="ABCD";
C. char * const p="ABCD";
D. const char * const p="ABCD";

【答】D

(16) 有以下程序段：

```
int i = 0, j = 1;
int &r = i;        //①
r = j;             //②
int * p = &i;      //③
* p = &r;          //④
```

会产生编译错误的语句是( )。

A. ④ B. ③ C. ② D. ①

【答】A

(17) 以下程序的输出结果是( )。

```
int main ()
{
    char * str = "12345";
    cout << strlen(str)<<"\t"<< sizeof(str);
    return 0;
}
```

A. 6 5 B. 5 6 C. 5 4 D. 5 5

【答】C

(18) 以下程序的输出结果是( )。

```
int main ()
{
    char w[][10] = { "ABCD", "EFGH", "IJKL", "MNOP"}, k;
    for (k = 1; k < 3; k++)    cout << w[k]<< endl;
    return 0;
}
```

A. ABCD B. ABCD C. EFG D. EFGH
   FGH   EFG   JK   IJKL
   KL   IJ   O
   M

【答】D

(19) 已知"char str1[8], str2[8] = {"good"};"，则在程序中不能将字符数组 str2 赋给 str1 的语句是（　　）。

  A. str1＝str2;　　　　　　　　B. strcpy (str1, str2);

  C. strncpy (str1, str2, 6);　　　　D. memcpy (str1, str2, 5);

【答】A

(20) 变量的指针是指该变量的（　　）。

  A. 值　　　　B. 地址　　　　C. 名　　　　D. 一个标志

【答】B

(21) 对于变量 p，sizeof(p) 的值不为 4 的是（　　）。

  A. short int p;　　　　　　　　B. char ＊＊＊ p;

  C. double ＊ p;　　　　　　　　D. char ＊ p＝"12345";

【答】A

(22) 下面能正确进行字符串赋值操作的是（　　）。

  A. char s[5] = {"ABCDE"};　　　B. char s[5] = {'A', 'B', 'C', 'D', 'E'};

  C. char ＊ s; s = "ABCDE";　　　D. char ＊ s;　cin >> s;

【答】C

(23) 对于指向同一块连续内存的两个指针变量不能进行的运算是（　　）。

  A. ＜　　　　B. ＝　　　　C. ＋　　　　D. －

【答】C

(24) 若有语句"int ＊ point, a = 4;"和"point = &a;"，下面均代表地址的一组选项是（　　）。

  A. a, point, ＊&a　　　　　　　B. &＊a, &a, ＊point

  C. ＊&point, ＊point, &a　　　　D. &a, &＊point, point

【答】D

(25) 已有定义"int k = 2; int ＊ ptr1, ＊ ptr2;"，且 ptr1 和 ptr2 均指向变量 k，下面不能正确执行的赋值语句是（　　）。

  A. k = ＊ptr1 ＋ ＊ptr2;　　　　B. ptr2 = k;

  C. ptr1 = ptr2;　　　　　　　　D. k = ＊ptr1 ＊ (＊ptr2);

【答】B

(26) 若有说明"int i, j = 2, ＊p = &i;"，则能完成 i = j 赋值功能的语句是（　　）。

  A. i = ＊p;　　B. ＊p = ＊&j;　　C. i = &j;　　D. i = ＊＊p;

【答】B

(27) 若有定义"int a[8];"，则以下表达式中不能代表数组元素 a[1] 的地址的是（　　）。

  A. &a[0] ＋ 1　　B. &a[1]　　C. &a[0]++　　D. a ＋ 1

【答】C

(28) 若有以下语句且 0≤k<6，则正确表示数组元素地址的表达式是（　　）。

  int x[] = {1, 3, 5, 7, 9, 11}, ＊ptr = x, k;

  A. x++　　　　B. &ptr　　　　C. &ptr[k]　　　　D. &(x ＋ 1)

【答】C

(29) 下面程序段的运行结果是(   )。

```
char *p = "abcdefgh";
p += 3;
cout << strlen(strcpy(p, "ABCD"));
```

  A. 8   B. 12   C. 4   D. 出错

【答】 D

【注解】 strcpy(p, "ABCD")出错,要求 p 为指向数组的指针,若改为:

```
char a[] = "abcdefgh";
char *p = a;
p += 3;
cout << strlen(strcpy(p, "ABCD"));
```

则结果为 4。

(30) 设有语句"int array[3][4];",则在下面几种引用下标为 i 和 j 的数组元素的方法中,不正确的引用方式是(   )。

  A. array[i][j]   B. *(*(array + i) + j)
  C. *(array[i] + j)   D. *(array + i*4+j)

【答】 D

(31) 若有以下定义和语句,则对 s 数组元素的正确引用形式是(   )。

```
int s[4][5], (*ps)[5];
ps = s;
```

  A. ps+1   B. *(ps+3)   C. ps[0][2]   D. *(ps+1)+3

【答】 C

【注解】 其他均为地址表达式。

(32) 在说明语句"int *f();"中,标识符 f 代表的是(   )。

  A. 一个用于指向整型数据的指针变量
  B. 一个用于指向一维数组的行指针
  C. 一个用于指向函数的指针变量
  D. 一个返回值为指针型的函数名

【答】 D

(33) 函数原型为 fun(int (*p)[3],int),调用形式为 fun(a,2),则 a 的定义应该为(   )。

  A. int **a B. int (*a)[] C. int a[][3] D. int a[3]

【答】 C

(34) 若"int i=100;",在下列引用方法中,正确的是(   )。

  A. int &r=i; B. int &r=100; C. int &r; D. int &r=&i;

【答】 A

(35) 在下列引用方法中,错误的是(   )。

  A. int i;   B. int i;   C. float f;   D. char c;
   int &r=i;   int &r; r=i;   float &r=f;   char &r=c;

【答】 B

(36) 以下程序的执行结果为（　　）。

```
int f(int i) {return ++i;}
int g(int &i) {return ++i;}
int main() {
    int a(0),b(0);
    a += f(g(a));
    b += f(f(b));
    cout << a <<"\t"<< b;
    return 0;
}
```

A. 3　2　　　　B. 2　3　　　　C. 3　3　　　　D. 2　2

【答】 A

(37) 以下程序的执行结果为（　　）。

```
int& max(int& x, int& y) { return (x > y?x:y); }
int main() {
    int m(3),n(4);
    max(m,n) -- ;
    cout << m <<"\t"<< n;
    return 0;
}
```

A. 3　2　　　　B. 2　3　　　　C. 3　4　　　　D. 3　3

【答】 D

(38) 若定义结构体：

```
struct st {
int no;
char name[15];
float score; } s1;
```

则结构体变量 s1 所占的内存空间为（　　）。

A. 15

B. sizeof(int)＋sizeof(char[15])＋sizeof(float)

C. sizeof(s1)

D. max(sizeof(int)，sizeof(char[15])，sizeof(float))

【答】 C

【注解】 对结构体所占的内存空间,避免使用成员空间相加的方法。

(39) 若定义联合体：

```
union { int no;
char name[15];
float score; } u1;
```

则联合体变量 u1 所占的内存空间为（　　）。

A. 15

B. sizeof(int)＋sizeof(char[15])＋sizeof(float)

C. sizeof(u1)

D. max(sizeof(int)，sizeof(char[15])，sizeof(float))

【答】 C

【注解】 对联合体所占的内存空间避免使用成员空间相加的方法。

(40) 当定义"const char * p="ABC";"时,下列语句正确的是(　　)。

  A. char * q=p; B. p[0]='B'; C. * p='\0'; D. p=NULL;

【答】 D

(41) 下列语句错误的是(　　)。

  A. const int a[4]={1,2,3}; B. const int a[]={1,2,3};

  C. const char a[3]={'1','2','3'}; D. const char a[]="123";

【答】 A

(42) 下列语句错误的是(　　)。

  A. const int buffer=256; B. const int temp;

  C. const double * point; D. const double * p=new double(3.1);

【答】 B

2. 程序填空题

(1) 以下是一个评分统计程序,共有 8 位评委打分,统计时,去掉一个最高分和一个最低分,其余 6 个分数的平均分即是最后得分,最后显示得分,显示精度为一位整数、两位小数。程序如下,请将程序补充完整。

```
#include <iostream>
using namespace std;
int main()
{
    float x[8] = ___①___ ;{0};
    float aver(0), max___②___ , min___③___ ;
    for (int i = 0; i < 8; i++) {
        cin >> x[i];
        if (x[i]> max)
            ___④___ ;
        if ( ___⑤___ )
            min = x[i];
        aver += x[i];
        cout << x[i]<< endl;
    }
    aver = ___⑥___ ;
    cout << aver << endl;
    return 0;
}
```

【答】 程序如下:

```
#include <iostream>
using namespace std;
int main()
{
    float x[8] = {0};
    float aver(0), max(0), min(200);
```

```cpp
    for (int i = 0; i < 8; i++) {
        cin >> x[i];
        if (x[i]> max)
            max = x[i];
        if (x[i]< min)
            min = x[i];
        aver += x[i];
        cout << x[i]<< endl;
    }
    aver = (aver - max - min)/6;
    cout << aver << endl;
    return 0;
}
```

(2) 以下程序在 M 行 N 列的二维数组中找出每一行上的最大值,显示最大值的行号、列号、值。请将程序补充完整。

```cpp
# include < iostream >
using namespace std;
int main()
{
    ____①____ ;
    int x[M][N] = {1, 5, 6, 4, 2, 7, 4, 3, 8, 2, 3, 1};
    for (____②____ ; i < M; i++)
    {
        int t = 0;
        for (____③____ ; j < N; j++)
            if (____④____)
                ____⑤____ ;
        cout << i + 1 <<","<< t + 1 <<" = "<< x[i][t]<< endl;
    }
    return 0;
}
```

【答】 程序如下:

```cpp
# include < iostream >
using namespace std;
int main()
{
    const int M(3),N(4);
    int x[M][N] = {1, 5, 6, 4, 2, 7, 4, 3, 8, 2, 3, 1};
    for (int i = 0; i < M; i++)
    {
        int t = 0;
        for ( int j = 0; j < N; j++)
            if (x[i][j]> x[i][t])
                t = j;
        cout << i + 1 <<","<< t + 1 <<" = "<< x[i][t]<< endl;
    }
```

```
        return 0;
    }
```

(3) 函数 expand(char *s, char *t)在将字符串 s 复制到字符串 t 时,将其中的换行符和制表符转换为可见的转义字符,即用"\n"表示换行符,用"\t"表示制表符。请填空。

```
void expand ( char * s, char * t)
{
for (int i = 0, int j = 0; s[i] != '\0'; i++)
    switch (s[i])
    {
        case '\n': t[   ①   ] = ___②___ ;
                   t[j++] = 'n';
                     ___③___ ;
        case '\t': t[   ④   ] = ___⑤___ ;
                   t[j++] = 't';
                   break;
        default:  t[   ⑥   ] = s[i];
                   break;
    }
    t[j] = ___⑦___ ;
}
```

【答】 程序如下:

```
void expand ( char * s, char * t)
{
for (int i = 0, int j = 0; s[i] != '\0'; i++)
    switch (s[i])
    {
        case '\n': t[ j++ ] = '\\';
                   t[j++] = 'n';
                   break;
        case '\t': t[ j++ ] = '\\';
                   t[j++] = 't';
                   break;
        default:   t[ j++ ] = s[i];
                   break;
    }
    t[j] = '\0';
}
```

3. 程序改错题

下列程序如有错请改正,并写出改正后的运行结果。

```
# include <iostream>
using namespace std;
int &add(int x, int y)
{
    return x + y;
}
int main()
```

```
    {
        int n(2),m(10);
        cout <<(add(n,m)  += 10) << endl;
        return 0;
    }
```

【答】 错误见横线处:

```
# include <iostream>
using namespace std;
int & add(int x, int y)
{
    return x + y;
}
int main()
{
    int n(2),m(10);
    cout <<(add(n,m)  += 10) << endl;
    return 0;
}
```

函数的类型是引用,因此返回的应为左值表达式。

修改后的正确程序如下:

```
# include <iostream>
using namespace std;
int & add(int &x, int y)
{
    x = x + y;
    return x;
}
void main()
{
    int n(2),m(10);
    cout <<(add(n,m)  += 10) << endl;
}
```

4. 编程题

(1) 输入 10 个整数,将这 10 个整数按升序排列输出,并且奇数在前、偶数在后。例如,如果输入的 10 个数是 10 9 8 7 6 5 4 3 2 1,则输出 1 3 5 7 9 2 4 6 8 10。

【答】 采用直观的"**选择排序法**"进行排序,基本步骤如下:

① 将 a[0] 依次与 a[1]~a[n−1] 比较,选出小者与 a[0] 交换,最后 a[0] 为 a[0]~a[n−1] 中的最小者;

② 将 a[1] 依次与 a[2]~a[n−1] 比较,选出小者与 a[1] 交换,最后 a[1] 为 a[1]~a[n−1] 中的最小者;

③ 同理,从 i=2 到 i=n−1,将 a[i] 依次与 a[i+1]~a[n−1] 比较,选出较小者存于 a[i] 中。

满足下列情况属于小者:

① 同为奇数或偶数,值较小者。
② 奇数、偶数中的奇数。
程序如下:

```cpp
#include <iostream>
using namespace std;
int main()
{
    const int MaxN = 10;
    int a[MaxN] = {10,9,8,7,6,5,4,3,2,1};
    for (int n = 0;n < MaxN;n++)
    {
        cin >> a[n];                    //输入数组元素
        if (a[n]< 0)
            break;
    }
    for (int i = 0;i < n;i++)
        cout << a[i]<<"\t";
    cout << endl <<"n = "<< n << endl;
    //对数组元素逐趟进行选择排序
    for (i = 0;i < n-1;i++)
        for (int j = i+1;j < n;j++)
            if ( (a[i]%2 == a[j]%2)&&a[i]> a[j] || a[i]%2 < a[j]%2 )
            {
                int t;
                t = a[i];               //交换数组元素
                a[i] = a[j];
                a[j] = t;
            }
    for (i = 0;i < n;i++)
        cout << a[i]<<"\t";              //显示排序结果
    return 0;
}
```

(2) 编程打印以下形式的杨辉三角形。

```
            1
           1 1
          1 2 1
         1 3 3 1
        1 4 6 4 1
       1 5 10 10 5 1
```

【答】 杨辉三角形有以下规律:
① 每行第一个数与最后一个数为 1。
② 每行其他数为前一行对应位置的前、后两数之和。
采用数组 a[i][j] 存储第 i 行 j 列的数:
① a[i][0]=1,a[i][i]=1; (i=0,1,2,…)

② a[i][j]=a[i-1][j-1]+a[i-1][j]; (j=1,2,3,…)

程序如下：

```
#include <iostream>
#include <iomanip>
using namespace std;
int main()
{
    const int MaxN = 10;
    int a[MaxN][MaxN];
    for (int i = 0; i < MaxN; i++) {
        a[i][0] = 1;
        a[i][i] = 1;
        for(int j = 1; j < i; j++)
            a[i][j] = a[i-1][j-1] + a[i-1][j];
    }
    for (i = 0; i < MaxN; i++) {            //显示结果
        cout << setw((MaxN - i) * 5/2);
        for(int j = 1; j <= i; j++)
            cout << a[i][j] << setw(5);
        cout << endl;
    }
    return 0;
}
```

(3) 编写一个程序，实现将用户输入的一个字符串以反向形式输出。例如，输入的字符串是 abcdefg，输出为 gfedcba。

【答】 程序如下：

```
#include <iostream>
using namespace std;
int main()
{
    char str[180];
    cin >> str;
    int k = strlen(str);
    for (int i = 0; i < k; i++)
        cout << str[k - i - 1];
    return 0;
}
```

(4) 编写一个程序，实现将用户输入的一个字符串中的所有字符'c'删除，并输出结果。

【答】 程序如下：

```
#include <iostream>
using namespace std;
int main()
{
```

```cpp
    char str[180];
    char c = 'c';
    cin >> str;
    for (int i = 0, j = 0; str[i]!= '\0';   i++, j++)
        if(str[j] == c)
            i-- ;
        else
            str[i] = str[j];
    cout << str << endl;
    return 0;
}
```

(5) 编写一个程序,将字符数组 s2 中的全部字符复制到字符数组 s1 中,不用 strcpy 函数。在复制时,'\0'也要复制过去,'\0'后面的字符不复制。

【答】 程序如下:

```cpp
#include <iostream>
using namespace std;
int main()
{
    char s2[180], s1[180];
    cin >> s2;
    for (int i = 0; s2[i]!= '\0';   i++)
        s1[i] = s2[i];
    s1[i] = s2[i];                    //复制'\0'
    cout << s1;
    return 0;
}
```

(6) 不用 strcat 函数编程实现字符串连接函数 strcat 的功能,将字符串 DStr 连接到字符串 SStr 的尾部。

【答】 若用静态数组存储字符串,SStr 必须有足够的空间,程序如下:

```cpp
#include <iostream>
using namespace std;
int main()
{
    char SStr[180], DStr[180];
    cin >> SStr >> DStr;
    int SL = strlen(SStr);
    int DL = strlen(DStr);
    for (int i = 0; i < DL + 1;   i++)
        SStr[i + SL] = DStr[i];
    cout << SStr;
    return 0;
}
```

(7) 编程求两个矩阵的乘积,要求两个矩阵的维数由键盘临时输入。

【答】 矩阵乘积的计算公式为 $c_{ij} = \sum_{k=1}^{K} a_{ik} \times b_{jk}$,其程序如下:

```cpp
#include <iostream>
using namespace std;
int main()
{
    const int M = 3, K = 2, N = 4;
    int a[M][K] = {{1,2},
                   {3,4},
                   {5,6}};
    int b[K][N] = {{1,0,1,1},
                   {0,1,0,1}};
    int c[M][N] = {0};
    for (int i = 0; i < M;   i++)
       for(int j = 0; j < N; j++)
          for(int k = 0; k < K; k++)
             c[i][j] = c[i][j] + a[i][k] * b[k][j];
    for (i = 0; i < M;   i++) {
        for(int j = 0; j < N; j++)
           cout << c[i][j]<<"\t";
        cout << endl;
    }
    return 0;
}
```

(8) 编写一个函数,判断输入的一串字符是否为"回文"。所谓"回文",是指顺读和倒读都一样的字符串。例如"level""ABCCBA"都是回文。

【答】 程序如下:

```cpp
#include <iostream>
using namespace std;
bool palindrome(char *str) {
    int h = strlen(str);
    for(int i = 0; i < h/2; i++)
        if (str[i]!= str[h-i-1])
            return false;
    return true;
}
int main()
{
    char text[180];
    cin >> text;
    if (palindrome(text))
        cout <<"a palindrome string!"<< endl;
    else
        cout <<"not a palindrome string!"<< endl;
    return 0;
}
```

(9) 编写一个函数 int SubStrNum (char * str, char * substr),它的功能是统计子字符串 substr 在字符串 str 中出现的次数。

【答】 程序如下：

```cpp
#include <iostream>
using namespace std;
int SubStrNum (char * str, char * substr){
    int Num = 0;                    //查找到子字符串的次数
    int h1,h2;                      //分别存储 str、substr 的长度
    int p1,p2;                      //分别指向 str、substr
    h1 = strlen(str);
    h2 = strlen(substr);
    p1 = 0;
    while(p1 < h1)                  // str 未到尾
    {
        p2 = 0;                     //重新开始比较
        while(str[p1] == substr[p2]&&p2 < h2&&p1 < h1)   //相等且均没到尾
        {
            p1++;
            p2++;                   //指针向后移
        }
        if(p1 == h1&&p2 < h2)       //如果 str 到尾但 substr 未到尾,结束比较
            break;
        if(p2 == h2)                //如果 substr 到尾,找到一个子串
            Num++;
        else                        // 有字符不相等,指针 p1 回移
            p1 = p1 - p2 + 1;
    }
    return Num;
}
int main() {
    char str[200];
    char substr[20];
    cout <<"input source string: ";
    cin >> str;
    cout <<"input sub string: ";
    cin >> substr;
    cout <<"match times:"<< SubStrNum (str, substr)<< endl;
    return 0;
}
```

(10) 编写一个函数,返回任意大的两整数之差(提示：大整数用字符串来表示)。

【答】 程序如下：

```cpp
#include <iostream>
using namespace std;
```

```cpp
char * lsub(char * s1, char * s2)
{
    int n1,n2,n;
    char * res,c = 0;
    n1 = strlen(s1);                    //n1 = 数字串 s1 的长度
    n2 = strlen(s2);                    //n2 = 数字串 s2 的长度
    n = n1 > n2 ? n1 : n2;              //数字串 s1、s2 的最大长度
    res = new char [n + 2];             //申请存结果串的内存
    res[0] = '0';                       //结果初值为 0
    res[1] = '\0';
    res[n + 1] = '\0';
    if(n1 > n2)
        res[0] = '+';                   //n1 > n2 结果为正
    else if(n1 < n2)
        res[0] = '-';                   //n1 < n2 结果为负
    else                                //n1 = n2
        for(int i = 0;i < n;i++) {
            if (s1[i]< s2[i]) {
                res[0] = '-';
                break;
            }
            if (s1[i]> s2[i]) {
                res[0] = '+';
                break;
            }
        }
    if(res[0] == '-') {
        for(int i = n; i > 0;i-- )      //将 s1 从低位开始搬到 res,没有数字的位填'0'
            res[i] = i > n-n1 ? s1[i+n1-n2-1] : '0';
        for(i=n; i>0; i--) {            //进行 s2 - res
            if(s2[i-1]-c >= res[i]) {                        //c 为上次相减的借位
                res[i] = s2[i-1]-c-res[i] + '0';
                c = 0;                  //本次相减借位
            }else {
                res[i] = s2[i-1] - res[i] - c + 10 + '0';
                c = 1;
            }
        }
    }
    if(res[0] == '+') {
        for(int i = n;i > 0;i-- )       //将 s2 从低位开始搬到 res,没有数字的位填'0'
            res[i] = i > n-n2 ? s2[i+n2-n1-1] : '0';
        for(i=n; i>0; i--) {            //进行 s1 - res
            if(s1[i-1]-c >= res[i]) {
                res[i] = s1[i-1] - res[i] - c + '0';
                c = 0;
            }else {
                res[i] = s1[i-1] - res[i] - c + 10 + '0';
                c = 1;
            }
        }
    }
```

```
        return res;
    }
    int main() {
        char num1[100],num2[100], * num;
        cin >> num1 >> num2;
        num = lsub(num1,num2);
        cout << num1 <<" - "<< num2 <<" = "<< num << endl;
        delete [] num;
        return 0;
    }
```

(11) 编写一个程序,求解约瑟夫(Josephus)问题。约瑟夫问题描述为:n 个小孩围成一圈做游戏,给定一个数 m,现从第 s 个小孩开始,顺时针计数,每数到 m,该小孩出列,然后从下一个小孩开始,当数到 m 时,该小孩出列,如此反复,直到最后一个小孩。

【答】 采用一个 n 结点的单向环形链表存储小孩编号:

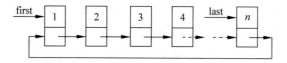

程序如下:

```
# include < iostream >
using namespace std;
struct Node
{
    int no;
    Node * next;
};
int main() {
    int m,n,s,remainder;              //remainder 表示剩下的小孩数
    cout <<"Input m,n,s:";
    cin >> m >> n >> s;
    Node * first, * last;
    first = last = new Node;
    first -> no = 1;
    for(int i = 1;i < n;i++) {        //构造链表
        Node * p = new Node;
        p -> no = i + 1;
        last -> next = p;
        last = p;
    }
    last -> next = first;             //围成环形链表
    for(i = 1;i < s;i++) {            //第 s 个孩子作为开始计数的小孩
        first = first -> next;
        last = last -> next;
    }
    remainder = n;
    while(remainder > 1) {
        for(int count = 1;count < m;count++) {
```

```cpp
            first = first->next;
            last = last->next;
        }
        last->next = first->next;    //第 m 个孩子出列
        cout << first->no <<"\t";
        delete first;
        first = last->next;
        remainder--;
    }
    cout << endl <<"The winner is No."<< first->no;
    delete first;
    return 0;
}
```

运行结果：

```
Input m,n,s:3 20 8↙
10    13    16    19    2    5    8    12    17
1     16    11    18    4    14   3    15    9
20
The winner is No.7
```

(12) 现有一个电话号码簿,其中有姓名、电话号码,当输入电话号码时,查找出姓名与电话号码;当输入姓名时,同样查找出姓名与电话号码;还允许不完全输入查找,例如输入 010 时查找出所有以 010 开头的号码,输入"杨"时列出所有姓名以"杨"开头的号码。

【答】 下列程序实现时功能有所加强,可以查出包含有输入号码或姓名的记录,而不仅仅是以输入号码与姓名为前缀的记录。

```cpp
#include <iostream>
using namespace std;
struct Phone
{
    char number[20];
    char name[16];
}PhoneBook[] = { {"027888888888",    "张建设"},
                {"010678923658",     "李涛"},
                {"0755628328178",    "王国维"},
                {"139012345608",     "杨振兴"},
                {"632478988",        "刘洋"},
                {"872537828",        "李欢"},
                {"972638288",        "慕容明昭"},
                {"0010678923658",    "Tom"}
};
bool SubStrNum (char *str, char *substr){
    int h1,h2;              //分别存储 str、substr 的长度
    int p1,p2;              //分别指向 str、substr
    h1 = strlen(str);
    h2 = strlen(substr);
```

```cpp
            if (h2 > h1)
                return false;
            p1 = 0;
            while(p1 < h1)                      //str 未到尾
            {
                p2 = 0;                         //重新开始比较
                while(str[p1] == substr[p2]&&p2 < h2&&p1 < h1)    //相等且均没到尾
                {
                    p1++;
                    p2++;                       //指针向后移
                }
                if(p1 == h1&&p2 < h2)           //如果 str 到尾但 substr 未到尾,结束比较
                    return false;
                if(p2 == h2)                    //如果 substr 到尾,找到一个子串
                    return true;
                else                            //有字符不相等,指针 p1 回移
                    p1 = p1 - p2 + 1;
            }
            return false;
}
int main(){
    int phones;
    char string[20];
    phones = sizeof(PhoneBook)/sizeof(Phone);              //电话记录条数
    cout <<"Input phone number or name:";
    cin >> string;
    for(int i = 0;i < phones;i++) {                         //在号码中找
        if(SubStrNum (PhoneBook[i].number, string))

    cout << i <<"\t"<< PhoneBook[i].number <<"\t"<< PhoneBook[i].name << endl;
    }
    for( i = 0;i < phones;i++) {                            //在名字中找
        if(SubStrNum (PhoneBook[i].name, string))

    cout << i <<"\t"<< PhoneBook[i].number <<"\t"<< PhoneBook[i].name << endl;
    }
    return 0;
}
```

运行结果：

```
Input phone number or name:010↙
1      010678923658       李涛
7      0010678923658      Tom
Input phone number or name:李↙
1      010678923658       李涛
5      872537828          李欢
```

1.4 习题 4 解答

1. 填空题

(1) 按变量的定义位置来分,变量可分为**全局变量**与**局部变量**。其中,**局部变量**定义在函数或复合语句中,供函数或复合语句中使用。

(2) 变量按存储类型分为 **auto**、**extern**、**register**、**static**,当声明一个**静态**(static)变量时,它既具有局部变量的性质,又具有全局变量的性质。

(3) C++程序的内存分为 4 个区,即**全局数据区**、**代码区**、**栈区**、**堆区**。全局变量、静态变量、字符串常量存放在**全局数据区**;所有的函数和代码存放在**代码区**;为运行函数而分配的函数参数、局部变量、返回地址存放在**栈区**;动态分配的内存在**堆区**。

(4) 全局变量、静态变量具有**静态**生存期;局部变量生存期为**动态**。

(5) 函数原型中形参标识符的作用域为**函数原型**,函数的形参与函数体作用域为**块作用域**;函数、全局变量与常量有**文件**作用域。

(6) C++源程序中以♯开头、以换行符结尾的行称为预处理命令。预处理命令**编译**前由预处理器执行。

(7) 用户可以通过 3 种方式使用名字空间,即**个别使用声明方式**、**全局声明方式**、**全局声明个别成员**。

2. 选择题

(1) 在 C++中,函数默认的存储类别为(   )。

　　A. auto　　　　　B. static　　　　　C. extern　　　　　D. 无存储类别

【答】 A

(2) 以下叙述不正确的是(   )。

　　A. 在不同的函数中可以定义相同名字的变量
　　B. 函数中的形式参数是局部变量
　　C. 在一个函数体内定义的变量只在本函数范围内有效
　　D. 在一个函数内的复合语句中定义的变量在本函数范围内有效

【答】 D

【注解】 在一个函数内的复合语句中定义的变量在复合语句中有效。

(3) 以下叙述不正确的是(   )。

　　A. 预处理命令行必须以♯号开头
　　B. 凡是以♯开头的语句都是预处理命令行
　　C. 在程序执行前执行预处理命令
　　D. ♯define PI=3.14 是一条正确的预处理命令

【答】 D

【注解】 正确的预处理命令应为♯define PI 3.14。

(4) 以下叙述不正确的是(   )。

　　A. 动态分配的内存要用 delete 释放
　　B. 局部 auto 变量分配的内存在函数调用结束时释放
　　C. 局部字符串常量、静态变量的内存在函数调用结束时释放

D. 全局变量的内存在程序结束时释放

【答】 C

(5) 下列程序段的运行结果为(　　)。

```
#define ADD(x) x + x
int main()
{
    int m = 1, n = 2, k = 3;
    int sum = ADD(m + n) * k;
    cout <<"sum = "<< sum;
    return 0;
}
```

A. sum＝9　　　　B. sum＝10　　　　C. sum＝12　　　　D. sum＝18

【答】 B

【注解】 sum ＝ ADD（m＋n）* k 替换成 sum＝m＋n＋m＋n * k。

(6) 下列程序段的运行结果为(　　)。

```
int i = 100;
int fun()
{
    static int i = 10;
    return ++i;
}
int main()
{
    fun();
    cout << fun()<<","<< i;
    return 0;
}
```

A. 10,100　　　　B. 12,100　　　　C. 12,12　　　　D. 11,100

【答】 B

【注解】 fun()中的 i 为静态变量,但只在 fun()中可见;main()中的 i 是全局变量,fun()对 i 的修改不影响 main()中的 i。

(7) 下列程序段的运行结果为(　　)。

```
#include <iostream>
using namespace std;
#define DEBUG 2
int main() {
    int i = 3;
#ifndef DEBUG
    for(,,)
        cout << DEBUG <<","<< i;
#else
        cout << i <<",";
#endif
        cout << DEBUG;
    return 0;
}
```

A. 2,3　　　　　　　　　　　　B. 2
C. 3,2　　　　　　　　　　　　D. 程序语句未写完,编译错

【答】 C

(8) 下列程序段的运行结果为(　　)。

```
#include<iostream>
using namespace std;
namespace mynamespace
{
    int flag = 10;
}
namespace yournamespace
{
    int flag = 100;
}
int main()
{
    int flag = 1000;
    using namespace yournamespace;
    cout<<flag<<","<<mynamespace::flag;
    return 0;
}
```

A. 100,1000　　　B. 1000,10　　　C. 1000,1000　　　D. 100,10

【答】 B

【注解】 "cout<<flag"中的 flag 在 yournamespace 的块外,因此为块外的 flag。

(9) 下列程序段的运行结果为(　　)。

```
char * inputa()
{
    char str[20] = "123";
    return str;
}
char * inputp()
{
    char * str = "123";
    return str;
}
int main()
{
    char * p = inputp();
    cout<<inputa()<<","<<p<<endl;
    return 0;
}
```

A. 乱码,123　　　B. 123,乱码　　　C. 123,123　　　D. 乱码,乱码

【答】 A

3. 程序填空题

为了使下列程序能顺利编译,请在空白处填上相应的内容:

```
#include ___①___
```

   ② int sumchar(char str[ ]);
int main( )
{
   ③ ;
   ④ ;
  char * p = new ⑤ ;
  cin >> p;
  cout << sumchar(p);
   ⑥ p;
  return 0;
}

【答】 程序如下:

```
#include <iostream>
extern int sumchar(char str[ ]);
int main( )
{
    using std::cin;
    using std::cout;
    char * p = new char [180];
    cin >> p;
    cout << sumchar(p);
    delete [ ]p;
    return 0;
}
```

4. 程序分析题(写出运行结果)

```
#include <iostream>
using namespace std;
const int PI = 3.14;                    //符号常量

const int * Fun( void )
{
    const int a = 5;                    //局部常变量
    const int * p = &a;                 //局部常指针变量
    cout << "Value of local const variable a: " << a << endl;
    cout << "Address of local const variable a: " << &a << endl;
    cout << "Value of local const pointer p: " << p << endl;
    cout << "Value of local const variable a: " << * p << endl << endl;
    return p;
}

int main( )
{
    const int * q;
    q = Fun();
    cout << "Main(): " << endl;
    cout << "Value of local const pointer q: " << q << endl;
    cout << "The return value of the function Func(): " << * q << endl;
```

```
    const char * str = "123ABC";        //字符串常量
    cout << "Address of string const  variable: " << (void *)str
        << endl;
    cout << "Value of string const  variablestr: " << str << endl;
    cout << "Address of global const variable PI: " << &PI << endl;
    cout << "Address of the function Fun(): " <<(void *)Fun << endl;
    return 0;
}
```

【答】 运行结果：

```
Value of local const variable a: 5
Address of local const variable a: 0018FEE8
Value of local const pointer p: 0018FEE8
Value of local const variable a: 5

Main():
Value of local const pointer q: 0018FEE8
The return value of the function Func(): 1638216
Address of string const variable: 0046C14C
Value of string constvariablestr: 123ABC
Address of global const variable PI: 0046C01C
Address of the function Fun(): 0040128F
```

【注解】 本程序主要测试常变量的使用。程序定义了局部常变量 a，它是一个局部变量，作用域仅限于函数 Fun()内部。其生存周期不是整个源程序，和普通变量一样，局限在函数 Fun()内部，当函数调用完成后，局部常变量 a 被释放，当将函数 Fun()的返回值赋给指针 q 后，输出指针 q 所指向的存储空间内存储的不再是变量 a 的值，而是 1638216（随机值）。程序中的指针常变量 p 指向局部常变量 a，所以指针 p 的值就是局部常变量 a 的地址。

1.5　习题 5 解答

1. 填空题

（1）类的**私有成员**只能被该类的成员函数或友元函数访问。
（2）类的数据成员不能在定义的时候初始化，而应该通过**构造函数**初始化。
（3）类成员默认的访问方式是 **private**。
（4）类的**公有成员函数的集合**是该类给外界提供的接口。
（5）类的**公有成员**可以被类作用域内的任何对象访问。
（6）为了能够访问某个类的私有成员，必须在该类中声明该类的**友元**。
（7）**类的静态成员**为该类的所有对象共享。
（8）每个对象都有一个指向自身的指针，称为 **this** 指针,通过使用它来确定其自身的地址。
（9）运算符 **new** 自动建立一个大小合适的对象并返回一个具有正确类型的指针。
（10）C++禁止**非 const 成员函数**访问 const 对象。
（11）在定义类的动态对象数组时，系统只能够自动调用该类的**构造**函数对其进行初始化。

(12) C++中语句"**const char * const** p="hello";"所定义的指针 p 和它所指的内容都不能被改变。

(13) 假定 AB 为一个类,则语句"**AB(AB& x);**"为该类拷贝构造函数的原型说明。

(14) 在 C++中,访问一个对象的成员所用的运算符是 **.** ,访问一个指针所指向对象的成员所用的运算符是 **—>**。

(15) 析构函数在对象的**生存期结束**时被自动调用,全局对象和静态对象的析构函数在**程序运行结束时**调用。

(16) 设 p 是指向一个类的动态对象的指针变量,则执行"delete p;"语句时将自动调用该类的**析构函数**。

2. 选择题

(1) 数据封装就是将一组数据和与这组数据有关的操作组装在一起,形成一个实体,这个实体也就是( )。

  A. 类   B. 对象   C. 函数体   D. 数据块

【答】 A

(2) 类的实例化是指( )。

  A. 定义类   B. 创建类的对象

  C. 指明具体类   D. 调用类的成员

【答】 B

(3) 已知 p 是一个指向类 Sample 数据成员 m 的指针,s 是类 Sample 中的一个对象。如果要给 m 赋值为 5,正确的是( )。

  A. s.p=5;   B. s—>p=5;   C. s.*p=5;   D. *s.p=5;

【答】 C

(4) 关于类和对象的说法不正确的是( )。

  A. 对象是类的一个实例

  B. 一个类只能有一个对象

  C. 一个对象只能属于一个具体的类

  D. 类与对象的关系和数据类型与变量的关系是相似的

【答】 B

(5) 下列说法错误的是( )。

  A. 封装是将一组数据和这组数据有关的操作组装在一起

  B. 封装使对象之间不需要确定的接口

  C. 封装要求对象具有明确的功能

  D. 封装使得一个对象可以像一个部件一样用在各种程序中

【答】 B

(6) 下面说法正确的是( )。

  A. 内联函数在运行时是将该函数的目标代码插入每个调用该函数的地方

  B. 内联函数在编译时是将该函数的目标代码插入每个调用该函数的地方

  C. 类的内联函数必须在类体内定义

  D. 类的内联函数必须在类体外通过加关键字 inline 定义

【答】 B

(7) 下列说法正确的是(　　)。

　　A. 类定义中只能说明函数成员的函数头,不能定义函数体

　　B. 类中的函数成员可以在类体中定义,也可以在类体之外定义

　　C. 类中的函数成员在类体之外定义时必须要与类声明在同一个文件中

　　D. 在类体之外定义的函数成员不能操作该类的私有数据成员

【答】B

(8) 下面关于对象概念的描述错误的是(　　)。

　　A. 对象就是C语言中的结构体变量

　　B. 对象代表着正在创建的系统中的一个实体

　　C. 对象是一个状态和操作(或方法)的封装体

　　D. 对象之间的信息传递是通过消息进行的

【答】A

(9) 在建立类的对象时(　　)。

　　A. 只为每个对象分配用于保存数据成员的内存

　　B. 只为每个对象分配用于保存函数成员的内存

　　C. 为所有对象的数据成员和函数成员分配一个共享的内存

　　D. 为每个对象的数据成员和函数成员同时分配不同内存

【答】A

(10) 有以下类定义：

```
class SAMPLE
{
    int n;
public:
    SAMPLE(int i = 0):n(i){}
    void setValue(int n0);
};
```

下列关于setValue成员函数的实现正确的是(　　)。

　　A. SAMPLE::setValue(int n0){n=n0;}

　　B. void SAMPLE::setValue(int n0){n=n0;}

　　C. void setValue(int n0){n=n0;}

　　D. setValue(int n0){n=n0;}

【答】B

(11) 在下面的类定义中,错误的语句是(　　)。

```
class sample
{
    public:
        sample(int val);        //①
        ~sample();              //②
    private:
        int a = 2.5;            //③
        sample();               //④
}
```

A. ①②③④ B. ② C. ③ D. ①②③

【答】 C

(12) 对于任意一个类,析构函数的个数最多为( )。

A. 0个 B. 1个 C. 2个 D. 3个

【答】 B

(13) 类的构造函数被自动调用执行的情况是在定义该类的( )时。

A. 成员函数 B. 数据成员
C. 对象 D. 友元函数

【答】 C

(14) 有关构造函数的说法不正确的是( )。

A. 构造函数的名字和类的名字一样
B. 构造函数在定义类变量时自动执行
C. 构造函数无任何函数类型
D. 构造函数有且只有一个

【答】 D

(15) ( )是析构函数的特征。

A. 一个类中只能定义一个析构函数
B. 析构函数名与类名没任何关系
C. 析构函数的定义只能在类体内
D. 析构函数可以有一个或多个参数

【答】 A

(16) 下列( )不是构造函数的特征。

A. 构造函数的函数名和类名相同
B. 构造函数可以重载
C. 构造函数可以设置默认参数
D. 构造函数必须指定类型说明

【答】 D

(17) 在下列函数原型中,可以作为类AA构造函数的是( )。

A. void AA(int); B. int AA();
C. AA(int)const; D. AA(int);

【答】 D

(18) 下列关于成员函数特征的描述,( )是错误的。

A. 成员函数一定是内联函数
B. 成员函数可以重载
C. 成员函数可以设置参数的默认值
D. 成员函数可以是静态的

【答】 A

(19) 不属于成员函数的是( )。

A. 静态成员函数 B. 友元函数
C. 构造函数 D. 析构函数

【答】 B

(20) 已知类 A 是类 B 的友元, 类 B 是类 C 的友元, 则( )。
　　A. 类 A 一定是类 C 的友元
　　B. 类 C 一定是类 A 的友元
　　C. 类 C 的成员函数可以访问类 B 的对象的任何成员
　　D. 类 A 的成员函数可以访问类 B 的对象的任何成员

【答】 D

(21) 关于动态存储分配, 下列说法正确的是( )。
　　A. new 和 delete 是 C++用于动态内存分配和释放的函数
　　B. 动态分配的内存空间也可以被初始化
　　C. 当系统内存不够时, 会自动回收不再使用的内存单元, 因此程序中不必用 delete 释放内存空间
　　D. 当动态分配内存失败时, 系统会立刻崩溃, 因此一定要慎用 new

【答】 B

(22) 静态成员函数没有( )。
　　A. 返回值　　　　　　　　　B. this 指针
　　C. 指针参数　　　　　　　　D. 返回类型

【答】 B

(23) 有以下类定义:

　　class Foo {int bar;};

则 Foo 类的成员 bar 是( )。
　　A. 公有数据成员　　　　　　B. 公有成员函数
　　C. 私有数据成员　　　　　　D. 私有成员函数

【答】 C

(24) 下列关于 this 指针的叙述正确的是( )。
　　A. 任何与类相关的函数都有 this 指针
　　B. 类的成员函数都有 this 指针
　　C. 类的友元函数都有 this 指针
　　D. 类的非静态成员函数才有 this 指针

【答】 D

(25) 下列程序的执行结果是( )。

```
# include <iostream>
using namespace std;
class Test {
public:
    Test(){ n += 2; }
    ~Test(){ n -= 3; }
    static int getNum() { return n; }
private:
    static int n;
};
```

```
int Test::n = 1;
int main()
{
    Test * p = new Test;
    delete p;
    cout << "n = " << Test::getNum() << endl;
    return 0;
}
```

  A. n=0   B. n=1   C. n=2   D. n=3

【答】A

(26) 下列程序执行后的输出结果是(   )。

```
#include <iostream>
using namespace std;
class AA{
    int n;
public:
    AA(int k):n(k){ }
    int get(){ return n;}
    int get()const{ return n+1;}
};
int main()
{
    AA a(5);
    const AA b(6);
    cout << a.get()<< b.get();
    return 0;
}
```

  A. 55   B. 57   C. 75   D. 77

【答】B

(27) 由于常对象不能被更新，因此(   )。

  A. 通过常对象只能调用它的常成员函数

  B. 通过常对象只能调用静态成员函数

  C. 常对象的成员都是常成员

  D. 通过常对象可以调用任何不改变对象值的成员函数

【答】A

(28) 有以下类定义：

```
class AA
{
    int a;
public:
    int getRef()const{return &a;}        //①
    int getValue()const{return a;}       //②
    void set(int n)const{a = n;}         //③
    friend void show(AA aa)const{cout << a;}  //④
};
```

其中的 4 个函数定义中正确的是（　　）。

　　A. ①　　　　B. ②　　　　C. ③　　　　D. ④

【答】B

(29) 有以下类定义：

```
class Point
{
    int x_,y_;
public:
    Point():x_(0),y_(0){}
    Point(int x,int y = 0):x_(x),y_(y){}
};
```

若执行语句：

　　Point a(2),b[3], * c[4];

则 Point 类的构造函数被调用的次数是（　　）。

　　A. 2 次　　　　B. 3 次　　　　C. 4 次　　　　D. 5 次

【答】C

(30) 有以下类定义：

```
class Test{
public:
    Test(){a = 0;c = 0;}                        //①
    int f(int a)const{ this -> a = a;}          //②
    static int g(){ return a;}                  //③
    void h(int b){Test::b = b;};                //④
private:
    int a;
    static int b;
    const int c;
};
int Test::b = 0;
```

在标注号码的行中，能被正确编译的是（　　）。

　　A. ①　　　　B. ②　　　　C. ③　　　　D. ④

【答】D

(31) 若有以下类声明：

```
class MyClass
{
public:
    MyClass(){ cout << 1;}
};
```

执行下列语句：

　　MyClass a,b[2], * P[2];

程序的输出结果是（　　）。

　　A. 11　　　　B. 111　　　　C. 1111　　　　D. 11111

【答】 B

(32) 有以下程序：

```
#include<iostream>
using namespace std;
class A
{
    public:
        static int a;
        void init(){a=1;}
        A(int a=2)
        {
            init();
            a++;
        }
};
int A::a=0;
A obj;
int main()
{
    cout<<obj.a;
    return 0;
}
```

运行时输出的结果是（　　）。

　　A. 0　　　　　　B. 1　　　　　　C. 2　　　　　　D. 3

【答】 B

(33) 有以下程序：

```
#include<iostream>
using namespace std;

class MyClass{
public:
    MyClass(){cout<<"A";}
    MyClass(char c)
    { cout<<c; }
    ~MyClass(){cout<<"B";}
};

int main()
{
    MyClass p1,*p2;
    p2=new MyClass('X');
    delete p2;
    return 0;
}
```

执行这个程序，计算机屏幕上将显示输出（　　）。

　　A. ABX　　　　B. ABXB　　　　C. AXB　　　　D. AXBB

【答】 D

3. 简答题

（1）C++中的空类默认产生哪些类成员函数？

【答】 对于一个空类，编译器默认产生4个成员函数，即默认构造函数、析构函数、拷贝构造函数和赋值函数。

（2）类和数据类型有何关联？

【答】 类相当于一种包含函数的自定义数据类型，它不占内存空间，是一个抽象的"虚"体，使用已定义的类建立对象就像用数据类型定义变量一样。对象在建立后，对象占据内存，变成了一个"实"体。类与对象的关系就像数据类型与变量的关系一样。其实，一个变量就是一个简单的不含成员函数的数据对象。

（3）类和对象的内存分配关系如何？

【答】 为节省内存，编译器在创建对象时只为各对象分配用于保存各对象数据成员初始化的值，并不为各对象的成员函数分配单独的内存空间，而是共享类的成员函数定义，即类中成员函数的定义为该类的所有对象所共享，这是C++编译器创建对象的一种方法，在实际应用中，我们仍要将对象理解为由数据成员和函数成员两部分组成。

（4）什么是浅拷贝？什么是深拷贝？二者有何异同？

【答】 构造函数用于建立对象时给对象赋初值以初始化新建立的对象。如果有一个现存的对象，在建立新对象时希望利用现存对象作为新对象的初值，即用一个已存在的对象去初始化一个新建立的对象。C++提供的拷贝构造函数用于在建立新对象时将已存在对象的数据成员值复制给新对象，以初始化新对象。拷贝构造函数在用类的一个对象去初始化该类的另一个对象时调用，以下3种情况相当于用一个已存在的对象去初始化新建立的对象，因此调用拷贝构造函数。

① 当用类的一个对象去初始化该类的另一个对象时。

② 如果函数的形参是类的对象，调用函数时，将对象作为函数实参传递给函数的形参时。

③ 如果函数的返回值是类的对象，函数执行完成，将返回值返回时。

原因在于默认的拷贝构造函数实现的只能是**浅拷贝**，即直接将原对象的数据成员值依次复制给新对象中对应的数据成员，并没有为新对象另外分配内存资源。这样，如果对象的数据成员是指针，两个指针对象实际上指向的是同一块内存空间。

当类的数据成员中有指针类型时，我们就必须定义一个特定的拷贝构造函数，该拷贝构造函数不仅可以实现原对象和新对象之间数据成员的复制，而且可以为新的对象分配单独的内存资源，这就是**深拷贝**构造函数。

（5）什么是this指针？它的作用是什么？

【答】 在一个类的成员函数中，有时希望引用调用它的对象，对此C++采用隐含的this指针来实现。**this指针**是一个系统预定义的特殊指针，指向当前对象，表示当前对象的地址。系统利用this指针明确指出成员函数当前操作的数据成员所属的对象。实际上，当一个对象调用其成员函数时，编译器先将该对象的地址赋给this指针，然后调用成员函数，这样成员函数对对象的数据成员进行操作时，就隐含使用了this指针。

一般而言，通常不直接使用this指针来引用对象成员，但在某些少数情况下可以使用this指针，例如重载某些运算符以实现对象的连续赋值等。

this指针不是调用对象的名称，而是指向调用对象的指针的名称。this的值不能改变，它总是指向当前调用对象。

(6) C++中的静态成员有何作用?它有何特点?

【答】 C++提供了静态成员,用于解决同一个类的不同对象之间数据成员和函数的共享问题。

静态成员的特点是不管这个类创建了多少个对象,其静态成员在内存中只保留一份副本,这个副本为该类的所有对象所共享。

面向对象方法中还有**类属性**(class attribute)的概念,类属性是描述类的所有对象的共同特征的一个数据项,对于任何对象实例,它的属性值是相同的,C++通过静态数据成员来实现类属性。

(7) 友元关系有何性质?

【答】 友元关系具有以下性质:

① 友元关系是不能传递的,不能被继承。如 B 类是 A 类的友元,C 类是 B 类的友元,C 类和 A 类之间如果没有声明就没有任何友元关系,不能进行数据共享。

② 友元关系是单向的,不具有交换性,如果声明 B 类是 A 类的友元,B 类的成员函数就可以访问 A 类的私有和保护数据,但 A 类的成员函数不能访问 B 类的私有和保护数据。

(8) 在 C++ 程序设计中,友元关系有什么优点和缺点?

【答】 友元概念的引入提高了数据的共享性,加强了函数与函数之间、类与类之间的相互联系,大大提高了程序的效率,这是友元的优点;但友元也破坏了数据隐蔽和数据封装,导致程序的可维护性变差,给程序的重用和扩充埋下了深深的隐患,这是友元的缺点。

(9) 如何实现不同对象的内存空间的分配和释放?

【答】 当类被实例化成对象后,不同类别的对象占据不同类型的内存,其规律与普通变量相同:

① 类的全局对象占有数据段的内存。

② 类的局部对象内存分配在栈中。

③ 函数调用时为实参建立的临时对象内存分配在栈中。

④ 使用动态内存分配语句 new 建立的动态对象,内存在堆中分配。

虽然类(对象)由数据成员和成员函数组成,但是,程序运行时,系统只为各对象的数据成员分配单独的内存空间,而该类的所有对象共享类的成员函数定义以及为成员函数分配的空间。对象的内存空间分配有下列规则:

① 对象的数据成员与成员函数占据不同的内存空间,数据成员的内存空间与对象的存储类别相关,成员函数的内存空间在代码段中。

② 一个类所有对象的数据成员拥有各自的内存空间。

③ 一个类所有对象的成员函数为该类的所有对象共享,在内存中只有一个副本。

随着对象的生命周期的结束,对象所占的空间就会释放,各类对象内存空间释放的时间与方法如下:

① 全局对象数据成员占有的内存空间在程序结束时释放。

② 局部对象与实参对象数据成员的内存空间在函数调用结束时释放。

③ 动态对象数据成员的内存空间要使用 delete 语句释放。

④ 对象的成员函数的内存空间在该类的所有对象生命周期结束时自动释放。

4. 程序填空题

(1) 在下面横线处填上适当字句,完成类中成员函数的定义。

```
class A{
    int * a;
public:
    A(int aa = 0) {
        a = _____①_____ ;                //用aa初始化a指向的动态对象
    }
    ~A(){_____②_____ ;};                //释放动态存储空间
```

【答】 ① new int(aa)
　　　② delete a

(2)

```
class A{
    _____①_____
    int n;
public:
    A(int nn = 0):n(nn){
        if(n == 0)a = 0;
        else a = new int[n];
    }
    _____②_____                //定义析构函数,释放动态数组空间
};
```

【答】 ① int * a;
　　　② ~A(){delete a;}

(3)

```
class Location {
    private:
        int X,Y;
    public:
        void init(int initX,int initY) {
            X = initX, Y = initY;
        }
        int GetX() {
            return X;
        }
        int GetY(){
            return Y;
        }
};
int main()
{
    Location A1; A1.init(20,90);
    _____①_____                //定义一个指向A1的引用rA1
    _____②_____                //用rA1在屏幕上输出对象A1的数据成员X和Y的值
    return 0;
}
```

【答】 ① Location & rA1=A1;
　　　② cout << rA1.GetX()<< rA1.GetY();

5. 程序分析题（写出运行结果）

（1）

```cpp
#include <iostream>
using namespace std;
class MyClass {
    public:
        int number;
        void set(int i);
};
int number = 3;
void MyClass::set (int i)
{
    number = i;
}
int main()
{
    MyClass my1;
    int number = 10;
    my1.set(5);
    cout << my1.number << endl;
    my1.set(number);
    cout << my1.number << endl;
    my1.set(::number);
    cout << my1.number;
    return 0;
}
```

【答】 运行结果：

```
5
10
3
```

（2）

```cpp
#include <iostream>
using namespace std;
class Location{
        public:
            int X,Y;
    void init (int initX,int initY)
    {
        X = initX, Y = initY;
    }
    int GetX()
    {
        return X;
    }
    int GetY()
    {
        return Y;
```

```
    }
};
void display(Location& rL)
{
    cout << rL.GetX()<<"   "<< rL.GetY()<<"\n";
}
int main()
{
    Location A[5] = {{0,0},{1,1},{2,2},{3,3},{4,4}};
    Location  * rA = A;
    A[3].init(5,3);
    rA -> init(7,8);
    for (int i = 0; i < 5; i++)
        display( * (rA++));
    return 0;
}
```

【答】 运行结果：

```
7   8
1   1
2   2
5   3
4   4
```

(3)

```
#include <iostream>
using namespace std;
class Test
{
    private:
        static int val;
        int a;
    public:
        static int func();
        void sfunc(Test &r);
};
int Test::val = 200;
int Test::func()
{
    return val++;
}
void Test::sfunc(Test &r)
{
    r.a = 125;
    cout <<"         Result3 = "<< r.a;
}
int main()
{
    cout <<"Result1 = "<< Test::func()<< endl;
    Test a;
```

```
        cout <<"Result2 = "<< a.func();
        a.sfunc(a);
        return 0;
}
```

【答】 运行结果：

```
Result1 = 200
Result2 = 201          Result3 = 125
```

（4）

```
#include <iostream>
using namespace std;
class Con
{
    char ID;
public:
    char getID()const{return ID;}
    Con():ID('A') {cout << 1;}
    Con(char ID):ID(ID){cout << 2;}
    Con(Con& c):ID(c.getID()) {cout << 3;}
};
void show(Con c)
{   cout << c.getID();}
int main()
{
    Con c1;
    show(c1);
    Con c2('B');
    show(c2);
    return 0;
}
```

【答】 运行结果：

```
13A23B
```

6. 改错题

（1）下面的程序有多处错误，说明错误原因并改正错误。

```
Class A {
      int a(0),b(0);
public:
      A(int aa,int bb) {a = aa;b = bb;}
}
A    x(2,3), y(4);
```

【答】 ① "int a(0),b(0);"错误，数据成员的初始化必须通过构造函数实现。

改正：int a,b;

② "A   x(2,3), y(4);"错误，对象 y 的初始化与已定义的构造函数参数表不匹配，可重

载一个构造函数对对象 y 进行初始化。

改正：A(int cc){a=cc;}

或：A x(2,3),y(4,5);

或：A(int aa,int bb=0) {a=aa;b=bb;}

(2) 下面的程序有一处错误，请用横线标出错误所在行并改正错误。

```
class Test{
    public;
        static int x;
};
int x = 20;                        //对类成员初始化
int main ()
{
    cout << Test::x;
    return 0;
}
```

【答】 "int x=20;"错误，因为 x 是静态数据成员，必须通过类名进行初始化。

改正：int Test::x=20;

(3) 用横线标出下面程序 main()函数中的错误行，并说明错误原因。

```
#include <iostream>
using namespace std;
class Location{
    private:
        int X,Y;
    public:
        void init(int initX,int initY) {
            X = initX;
            Y = initY;
        }
        int sumXY() {
            return X + Y;
        }
};
int main()
{
    Location A1;
    int x,y;
    A1.init(5,3);
    x = A1.X;y = A1.Y;
    cout << x + y <<"  "<< A1.sumXY()<< endl;
    return 0;
}
```

【答】 "x=A1.X; y=A1.Y;"错误，不能通过对象直接访问类的私有数据成员，可以改为公有数据成员或通过定义类的成员函数来实现。

改正：public:
        int X,Y;

（4）指出下面程序中的错误，并说明出错原因。

```cpp
#include <iostream>
using namespace std;
class ConstFun{
        public:
            void ConstFun(){}
            const int f5()const{return 5;}
            int Obj() {return 45;}
            int val;
            int f8();
};
int ConstFun::f8(){return val;}
int main()
{
        const ConstFun s;
        int i = s.f5();
        cout <<"Value = "<< endl;
        return 0;
}
```

【答】 "void ConstFun(){} ;"错误，构造函数不能有返回值。

改正：ConstFun(){} ；

7. 编程题

（1）定义一个三角形类 Ctriangle，求三角形的面积和周长。

【答】 程序如下：

```cpp
#include <iostream>
#include <cmath>
using namespace std;
class Ctriangle                         //定义三角形类 Ctriangle
{
public:
    Ctriangle(double x,double y, double z)
    {
        a = x;
        b = y;
        c = z;
    }
    double GetPerimeter()               //求三角形的周长
    {
        return a + b + c;
    }
    double GetArea()                    //求三角形的面积
    {
        double p = GetPerimeter()/2;
        return sqrt(p * (p - a) * (p - b) * (p - c));
    }
    void display()
    {
        cout <<"Ctriangle: "<<"a = "<< a <<" "<<"b = "<< b <<" "<<"c = "<< c << endl;
```

```cpp
        cout <<"Perimeter:"<< GetPerimeter()<< endl;
        cout <<"Area:"<< GetArea()<< endl;
    }
private:
    double a;
    double b;
    double c;
};
int main()
{
    Ctriangle T(3,4,5);
    T.display();
    return 0;
}
```

运行结果：

```
Ctriangle: a = 3 b = 4 c = 5
Perimeter:12
Area:6
```

【注解】 可以用顶点坐标定义类 Ctriangle。

(2) 定义一个点类 Point，并定义成员函数 double Distance(const& Point)求两点的距离。

【答】 程序如下：

```cpp
#include <iostream>
#include <cmath>
using namespace std;
class Point                          //定义类 Point
{
public:
    Point(double xx,double yy)
    {
        x = xx;
        y = yy;
    }
    double Getx()
    {
        return x;
    }
    double Gety()
    {
        return y;
    }
    double Distance(const Point &p)    //求两点之间的距离
    {
        x -= p.x;
        y -= p.y;
        return sqrt(x * x + y * y);
```

```cpp
    }
private:
    double x;
    double y;
};
int main()
{
    Point A(1,2),B(3,4);
    cout << A.Distance(B)<< endl;
    return 0;
}
```

运行结果：

```
2.82843
```

（3）定义一个日期类 Date，它能表示年、月、日。设计一个 NewDay()成员函数，增加一天日期。

【答】 程序如下：

```cpp
#include <iostream>
using namespace std;
class Date                              //定义日期类 Date
{
public:
    Date(int y = 2019, int m = 1, int d = 1);
    int days(int year, int month);
    void NewDay();
    void display()
    {   cout << year <<" - "<< month <<" - "<< day << endl; }
private:
    int year;                           //年
    int month;                          //月
    int day;                            //日
};

Date::Date(int y, int m, int d)
{
    if(m > 12 || m < 1){
        cout <<"Invalid month! "<< endl;
        m = 1;
    }
    if(d > days(y,m)) {
        cout <<"Invalid day! "<< endl;
        d = 1;
    }
    day = d;
    year = y;
```

```cpp
        month = m;
}

int Date::days(int year, int month)
{
    bool leap;
        if((year % 400 == 0)||(year % 4 == 0&&year % 100!= 0))
            leap = true;
    else
        leap = false;
    switch(month)
    {
        case 1:
        case 3:
        case 5:
        case 7:
        case 8:
        case 10:
        case 12:
            return 31;
        case 4:
        case 6:
        case 9:
        case 11:
            return 30;
        case 2:
            if(leap)                        //是否闰月
                return 29;
            else
                return 28;
            break;
    }
}

void Date::NewDay()
{
    if (day < days(year,month))
        day++;
    else {
        day = 1;
        month++;
        if(month == 13) {
            year++;
            month = 1;
        }
    }
}

int main()
{
    Date D1(2019,2,28);
    D1.display();
```

```
        D1.NewDay();
        cout <<"after a day:";
        D1.display();
        Date D2(2019,12,31);
        D2.display();
        D2.NewDay();
        cout <<"after a day:";
        D2.display();
        return 0;
    }
```

运行结果:

```
2019 - 2 - 28
after a day:2019 - 3 - 1
2019 - 12 - 31
after a day:2020 - 1 - 1
```

(4) 定义一个时钟类 Clock,设计成员函数 SetAlarm(int hour, int minute, int second) 设置响铃时刻;用 run()成员函数模拟时钟运行,当运行到响铃时刻时提示响铃。

【答】 程序如下:

```
# include < iostream >
using namespace std;
class Clock                                //定义时钟类 Clock
{
public:
    Clock(int h, int m, int s)
    {
        hour = (h > 23?0:h);
        minute = (m > 59?0:m);
        second = (s > 59?0:s);
    }
    void SetAlarm(int h, int m, int s )    //设置响铃时刻
    {
        Ahour = (h > 23?0:h);
        Aminute = (m > 59?0:m);
        Asecond = (s > 59?0:s);

    }
    void ShowTime()                         //显示时间
    {
        cout <<"Now: "<< hour <<":"<< minute <<":"<< second << endl;
    }
    void run()                              //时钟运行
    {
        second = second + 1;
        if(second > 59)
        {
```

```cpp
            second = 0;
            minute = minute + 1;
        }
        if(minute > 59)
        {
            minute = 0;
            hour = hour + 1;
        }
        if(hour > 23)
            hour = 0;
        //到响铃时刻
        if(hour == Ahour && minute == Aminute && second == Asecond)
        {
            cout <<"Plink!plink!plink!..."<< endl;      //响铃
        }
    }
private:
    int hour;                                           //时
    int minute;                                         //分
    int second;                                         //秒

    int Ahour;                                          //时(响铃)
    int Aminute;                                        //分(响铃)
    int Asecond;                                        //秒(响铃)
};
int main()
{
    Clock D1(7,59,57);
    D1.ShowTime();
    D1.SetAlarm(8,0,0);
    for(int i = 0;i < 3600 * 24 * 3 + 100;i++)
        D1.run();
    D1.ShowTime();
    return 0;
}
```

运行结果：

```
Now: 7:59:57
Plink!plink!plink!...
Plink!plink!plink!...
Plink!plink!plink!...
Now: 5:1:37
```

（5）设计一个学生类，包含学生学号、姓名、课程、成绩等基本信息，计算学生的平均成绩。

【答】 程序如下：

```cpp
# include < iostream >
# include < string >
```

```cpp
using namespace std;

class Student                              //定义学生类 Student
{
public:
    Student(char ID[ ],char name[ ],double g1,double g2,double g3)
    {
        num++;
        strcpy(this -> ID,ID);
        strcpy(this -> name,name);
        grade1 = g1;
        grade2 = g2;
        grade3 = g3;
        sum1 = sum1 + g1;
        sum2 = sum2 + g2;
        sum3 = sum3 + g3;
    }
    void display()
    {
        cout << ID <<"\t"<< name <<"\t"<< grade1 <<"\t"<< grade2 <<"\t"<< grade3
            << endl;
    }
    double average1()
    {
        return sum1/num;
    }
    double average2()
    {
        return sum2/num;
    }
    double average3()
    {
        return sum3/num;
    }
private:
    char    ID[10];                        //学号
    char    name[12];                      //姓名
    double grade1;                         //课程 1 成绩
    double grade2;                         //课程 2 成绩
    double grade3;                         //课程 3 成绩

    static double sum1;                    //课程 1 总分
    static double sum2;                    //课程 2 总分
    static double sum3;                    //课程 3 总分

    static int num;                        //学生总人数
};

int Student::num = 0;
double Student::sum1 = 0;
double Student::sum2 = 0;
double Student::sum3 = 0;
```

```cpp
int main()
{
    Student stu1("202006264","Li Weiwei",88,75,91);
    stu1.display();
    Student stu2("202002164","Chen Hanfu",86,78,93);
    stu2.display();
    Student stu3("202008079","Zhan Gaolin",94,69,97);
    stu3.display();

    cout <<"The average grade of course1: "<< stu1.average1()<< endl;
    cout <<"The average grade of course2: "<< stu2.average2()<< endl;
    cout <<"The average grade of course3: "<< stu2.average3()<< endl;
    return 0;
}
```

运行结果：

```
202006264        Li Weiwei        88        75        91
202002164        Chen Hanfu       86        78        93
202008079        Zhan Gaolin      94        69        97
The average grade of course1: 89.3333
The average grade of course2: 74
The average grade of course3: 93.6667
```

（6）有一个信息管理系统，要求检查每一个登录系统的用户(User)的用户名和口令，系统检查合格以后方可登录系统，用C++程序予以描述。

【答】 程序如下：

```cpp
#include <iostream>
#include <string>
using namespace std;
const int N = 20;
class User                                      //定义用户类 User
{
public:
    User(char * name, char * pass)
    {
        strcpy(username[num],name);
        strcpy(password[num],pass);
        for(int i = 0;password[num][i]!= '\0';i++)     //口令加密
            password[num][i] += i;
        num++;
    }
    void Adduser(char * name, char * pass)
    {
        strcpy(username[num],name);
        strcpy(password[num],pass);
```

```cpp
        for(int i = 0;password[num][i]!= '\0';i++)     //口令加密
            password[num][i] += i;
        num++;
    }
    int login(char * name, char * pass)
    {
        for(int i = 0;i < num;i++)
            if(strcmp(username[i],name) == 0)
            {
                for(int j = 0;pass[j]!= '\0';j++)
                    if(password[i][j]!= pass[j] + j)//口令解密
                        return -1;
                return i;
            }
        return -1;
    }
    private:
    char username[N][10];
    char password[N][10];
    static int num;
};
int User::num = 0;
int main()
{
    char name[10],pass[10];
    User u1("LiWei","liwei99");
    u1.Adduser("ChenHanfu","20200208");
    u1.Adduser("ZhanGaolin","199146");
    cout <<"Input username:";
    cin >> name;
    cout <<"Input password:";
    cin >> pass;
    if(u1.login(name, pass)> = 0)
        cout <<"Success login!"<< endl;
    else
        cout <<"login fail!"<< endl;
    return 0;
}
```

运行结果：

```
Input username:ZhanGaolin ↙
Input password:199146 ↙
Success login!
```

(7) 定义一个字符串类 String,增加下列成员函数。
- bool IsSubstring(const char * str)：判断 str 是否为当前串的子串；
- bool IsSubstring(const String & Str)：判断 str 是否为当前串的子串；
- int str2num()：把数字串转换成数；

- void toUppercase()：把字符串转换成大写字母。

【答】 程序如下：

```cpp
#include <iostream>
using namespace std;

class String
{
private:
    char * mystr;                                    //字符串
    int len;                                         //字符串长度
public:
    String()
    {
        len = 0;
        mystr = NULL;
    }
    String(const char * p)
    {
        len = strlen(p);
        mystr = new char[len + 1];
        strcpy(mystr, p);
    }
    String(String &r)                                //定义拷贝构造函数,防止浅拷贝
    {
        len = r.len;
        if (len != 0)
        {
            mystr = new char[len + 1];
            strcpy(mystr, r.mystr);
        }
    }
    ~String()
    {
        if (mystr != NULL)
        {
            delete[]mystr;
            mystr = NULL;
            len = 0;
        }
    }

    char * GetStr()const {return mystr;}
    void ShowStr()const {cout << mystr;}

    bool IsSubstring(const char * str);              //判断 str 是否为当前串的子串
    bool IsSubstring(const String &str);             //判断 String 类对象 str 是否
                                                     //为当前串的子串
    int str2num();                                   //把数字串转换成数
    void toUppercase();                              //把字符串转换成大写字母
};
```

```cpp
bool String::IsSubstring(const char * str)          //判断str是否为当前串的子串
{
    int i,j;
    for (i = 0;i < len;i++)
    {
        int  k = i;
        for (j = 0;str[j];j++,k++)
            if (str[j]!= mystr[k]) break;
        if (!str[j])   return true;
    }
    return false;
}

bool String::IsSubstring(const String &str)
{
    char * pstr = str.mystr;
    int plen = str.len;
    int i,j;
    for (i = 0;i < len;i++)
    {
        int k = i;
        for (j = 0;j < plen;j++,k++)
            if (pstr[j]!= mystr[k])  break;
        if (j == plen)   return true;
    }
    return false;
}

int String::str2num()                               //把数字串转换成数
{
    int num = 0;
    int i = -1;
    while (mystr[++i]!= 0)
        if (mystr[i]>= '0'&&mystr[i]<= '9')
            num = num * 10 + mystr[i] - '0';
    return num;
}

void String::toUppercase()                          //把字符串转换成大写字母
{
    int i = -1;
    while (mystr[++i]!= 0)
        if (mystr[i]>= 'a'&&mystr[i]<= 'z')
            mystr[i] = mystr[i] - 'a' + 'A';
}

int main()
{
    String s("abcdefaaaghijklmnopqrst");            //初始化字符串类对象
    String s1(s);                                   //调用拷贝构造函数
    cout <<"s1 = "<< s1.GetStr()<< endl;            //显示第一个字符串
    char substr1[30];
```

```cpp
        cout <<"Please input a test string:"<< endl;
        cin >> substr1;                                    //输入一个字符串,判断是否为 s1 的子串
        cout << substr1 <<" is "<< s1.GetStr()<<" substring?   ";
        cout <<((s1.IsSubstring (substr1))?"YES":"NO")<< endl;
        String substr2("CDEF");
        cout << substr2.GetStr()<<" is "<< s1.GetStr()<<" substring?   ";
        cout <<((s1.IsSubstring (substr2))?"YES":"NO")<< endl;
        s1.toUppercase();                                  //测试转换大写函数
        cout << substr2.GetStr()<<" is "<< s1.GetStr()<<" substring?   ";
        cout <<((s1.IsSubstring (substr2))?"YES":"NO")<< endl;
        String s2("123456 ");                              //测试将字符串转换成数字函数
        cout <<"Transform a string to a number:"<< endl;
        s2.ShowStr();
        int d = s2.str2num();
        cout <<"The Number is "<< d << endl;
        return 0;
    }
```

运行结果:

```
s1 = abcdefaaaghijklmnopqrst
Please input a test string:
rst ↙
rst is abcdefaaaghijklmnopqrst substring?    YES
CDEF is abcdefaaaghijklmnopqrst substring?   NO
CDEF is ABCDEFAAAGHIJKLMNOPQRST substring?   YES
Transform a string to a number:
123456 The Number is 123456
```

(8) 定义一个元素类型为 int、元素个数不受限制的集合类 Set。除了定义一些必要的函数外,还定义具有下列功能的成员函数。

- bool IsEmpty():判断集合是否为空;
- int size():返回元素个数;
- bool IsElement(int e) const:判断 e 是否属于集合;
- bool IsSubset(const Set& s)const:判断 s 是否包含于集合;
- bool IsEqual(const Set& s)const:判断集合是否相等;
- Set& insert(int e):将 e 加入到集合中;
- Set union(const Set& s) const:求集合的并;
- Set intersection(const Set& s) const:求集合的交;
- Set difference(const Set& s) const:求集合的差。

【答】 程序如下:

```cpp
#include <iostream>
using namespace std;
class Set
{
private:
```

```cpp
        int n;                                  //元素个数
        int *pS;                                 //集合
public:
    Set()    {n = 0;pS = NULL;}                  //符合习惯,集合下标从1开始有效
    Set(Set &s)                                  //用一个已存在的集合来初始化拷贝构造函数
    {
        n = s.n;
        if (n!= 0)
        {
            pS = new int[n + 1];
            for (int i = 1;i <= n;i++) pS[i] = s.pS[i];
        }
    }
    ~Set()                                       //析构函数
    {
        if (pS)
        {
            delete []pS;
            pS = NULL;
            n = 0;
        }
    }
    void ShowElement()const                      //显示集合元素
    {
        cout <<"{ ";
        for (int i = 1;i < n;i++)
            cout << pS[i]<<", ";
        cout << pS[n]<<" }"<< endl;
    }
    bool IsEmpt() {return n?false:true;}         //判断集合是否为空
    int size() {return n;}                       //返回元素个数
    bool IsElement(int e)const;                  //判断e是否属于集合
    bool IsSubset(const Set &s)const;            //判断s是否包含于集合
    bool IsEqual(const Set &s)const;             //判断集合是否相等
    Set & insert(int e);                         //将e加入到集合中
    Set Union(const Set& s)const;                //求集合的并
    Set intersection(const Set &s)const;         //求集合的交
    Set difference(const Set &s)const;           //求集合的差
};
bool Set::IsElement(int e)const                  //判断e是否属于集合
{
    for (int i = 1;i <= n;i++)
        if (pS[i] == e)
            return true;
    return false;
}
bool Set::IsSubset(const Set &s)const            //判断s是否包含于集合
{
    if (s.n > n)
        return false;
    for (int i = 1;i <= s.n;i++)                 //逐个判断集合s中的元素是否属于当前集合
        if (!IsElement(s.pS[i]))
```

```cpp
            return false;
    return true;
}
bool Set::IsEqual(const Set &s)const          //判断集合是否相等
{
    if (n!= s.n)
        return false;
    if (IsSubset(s))            //如果两集合元素个数相等,而且有一个为另一个的子集,则两集合相等
        return true;
    return false;
}
Set & Set::insert(int e)                      //将 e 加入到集合中
{
    if (IsElement(e))
        return * this;                        //若元素已存在则直接返回
    Set tempe;
    for (int i = 1;i <= n;i++)
        tempe.insert(pS[i]);
    n++;
    if (pS!= NULL)
        delete[ ]pS;
    pS = new int[n + 1];                      //重新申请
    for (i = 1;i < n;i++)
        pS[i] = tempe.pS[i];                  //复制
    pS[n] = e;                                //插入
    return * this;
}
//拷贝构造函数不能在常成员函数中使用
Set Set::Union(const Set &s)const             //求集合的并
{
    Set Rs;
    for (int i = 1;i <= n;i++)
        Rs.insert(pS[i]);                     //先用当前的集合初始化结果集
    for(i = 1;i <= s.n;i++)
    {
        if (!IsElement(s.pS[i]))
            Rs.insert(s.pS[i]);
    }
    return Rs;
}
Set Set::intersection(const Set &s)const      //求集合的交
{
    Set Rs;
    for (int i = 1;i <= s.n;i++)              //集合 s 中的元素是否属于当前集合,若属于则加入
        if (IsElement(s.pS[i]))
            Rs.insert(s.pS[i]);
    return Rs;
}
Set Set::difference(const Set &s)const        //求集合的差 ( * this) - s
{
    Set Rs;
    for (int i = 1;i <= n;i++)
```

```cpp
            if (!(s.IsElement(pS[i])))
                Rs.insert(pS[i]);
    return Rs;
}

int main()
{
    Set s1;                         //生成一集合 s1
    s1.insert(1);                   //测试插入函数,判断集合元素是否属于函数
    if (!s1.IsElement(3))
        s1.insert(3);
    s1.insert(1); s1.insert(2);

    cout <<"s1 = ";
    s1.ShowElement();               //测试判集合空函数
    cout <<"size:"<< s1.size ()<< endl;

    Set s2;                         //生成一个和集合 s1 不同的集合 s2
    s2.insert(1); s2.insert(3);s2.insert(5); s2.insert(6);
    cout <<"s2 = ";
    s2.ShowElement();

    Set s3(s1);                     //调用拷贝构造函数,生成一个和集合 s1 相同的集合 s3
    cout <<"s3 = ";
    s3.ShowElement();

    cout <<"s1 == s2? ";            //测试判集合相等函数
    cout <<((s1.IsEqual(s2))?"YES":"NO")<< endl;
    cout <<"s1 == s3? ";
    cout <<((s1.IsEqual(s3))?"YES":"NO")<< endl;

    Set s4;                         //生成一个为集合 s1 子集的集合,测试判子集 IsSubset 函数
    s4.insert(1);   s4.insert(3);
    cout <<"s4 = ";
    s4.ShowElement();

    cout <<"s4 is a subset of s1? ";
    cout <<((s1.IsSubset(s4))?"YES":"NO")<< endl;
    //测试集合并函数
    cout <<"s1 Union s2 = ";
    s1.Union(s2).ShowElement();
    //测试集合交函数
    cout <<"s1 intersection s2 = ";
    s1.intersection(s2).ShowElement();
    //测试集合差函数
    cout <<"s1 difference s2 = ";
    s1.difference(s2).ShowElement();
    return 0;
}
```

运行结果：

```
s1 = { 1, 3, 2 }
size:3
s2 = { 1, 3, 5, 6 }
s3 = { 1, 3, 2 }
s1 == s2? NO
s1 == s3? YES
s4 = { 1, 3 }
s4 is a subset of s1? YES
s1 Union s2 = { 1, 3, 2, 5, 6 }
s1 intersection s2 = { 1, 3 }
s1 difference s2 = { 2 }
```

1.6 习题 6 解答

1．填空题

（1）C++程序设计的关键之一是利用**继承**实现软件重用，有效地缩短程序的开发时间。

（2）派生类的对象可以作为基类的对象使用，这称为**类型兼容（或赋值兼容）**。

（3）在 C++中，3 种派生方式的说明符号为 **public**、**private**、**protected**，如果不加说明，则默认派生方式为 **private**。

（4）当私有派生时，基类的公有成员成为派生类的**私有成员**；保护成员成为派生类的**私有成员**；私有成员成为派生类的**不可访问的成员**。

（5）相互关联的各个类之间的关系主要分为**组合**关系和**继承**关系。

（6）在派生类中不能直接访问基类的**私有成员**，否则破坏了基类的封装性。

（7）保护成员具有双重角色，对派生类的成员函数而言，它是**公有成员**，但对所在类之外定义的其他函数而言则是**私有成员**。

（8）在多继承时，多个基类中的同名成员在派生类中由于标识符不唯一而出现**二义性**。在派生类中通过采用**成员名限定**或**重定义具有二义性的成员**来消除该问题。

（9）C++提供的**多继承**机制允许一个派生类继承多个基类。

2．选择题

（1）下面对派生类的描述错误的是（　　）。

  A．一个派生类可以作为另外一个派生类的基类

  B．派生类至少有一个基类

  C．派生类的成员除了它自己的成员外，还包含了它的基类成员

  D．派生类中继承的基类成员的访问权限在派生类中保持不变

【答】　D

（2）在多继承中，公有派生和私有派生对于基类成员在派生类中的可访问性与单继承的规则是（　　）。

  A．完全相同　　　　　　　　　　　　B．完全不同

  C．部分相同，部分不同　　　　　　　　D．以上都不对

【答】　A

(3) 下列对友元关系的叙述正确的是（　　）。
    A. 不能继承
    B. 是类与类的关系
    C. 是一个类的成员函数与另一个类的关系
    D. 提高程序的运行效率
【答】A

(4) 下面叙述不正确的是（　　）。
    A. 为了充分利用现有类，派生类一般都是公有派生
    B. 对基类成员的访问必须是无二义性的
    C. 赋值兼容规则也适用于多重继承的场合
    D. 基类的公有成员在派生类中仍然是公有的
【答】D

(5) 下面叙述不正确的是（　　）。
    A. 基类的保护成员在派生类中仍然是保护的
    B. 基类的保护成员在公有派生类中仍然是保护的
    C. 基类的保护成员在私有派生类中仍然是私有的
    D. 对基类成员的访问必须是无二义性的
【答】A

(6) 在公有派生的情况下，派生类中定义的成员函数只能访问原基类的（　　）。
    A. 公有成员和私有成员
    B. 私有成员和保护成员
    C. 公有成员和保护成员
    D. 私有成员、保护成员和公有成员
【答】C

(7) 一个类可以同时继承多个类，称为多继承，下列关于多继承和虚基类的表述错误的是（　　）。
    A. 每个派生类的构造函数都要为虚基类构造函数提供实参
    B. 多继承时有可能出现对基类成员访问的二义性问题
    C. 使用虚基数类可以解决二义性问题并实现运行时的多态性
    D. 建立派生类对象时，虚基数的构造函数会首先被调用
【答】C

(8) 在一个派生类对象结束其生命周期时（　　）。
    A. 先调用派生类的析构函数，后调用基类的析构函数
    B. 先调用基类的析构函数，后调用派生类的析构函数
    C. 如果基数没有定义析构函数，则只调用派生类的析构函数
    D. 如果派生类没有定义析构函数，则只调用基类的析构函数
【答】A

(9) 当保护继承时，基类的（　　）在派生类中成为保护成员，不能通过派生类的对象直接访问。
    A. 任何成员　　　　　　　　　　　　B. 公有成员和保护成员

C. 公有成员和私有成员　　　　　　D. 私有成员

【答】 B

(10) 若派生类的成员函数不能直接访问基类中继承来的某个成员，则该成员一定是基类中的(　　)。

　　A. 私有成员　　　　　　　　　　B. 公有成员
　　C. 保护成员　　　　　　　　　　D. 保护成员或私有成员

【答】 A

(11) 设置虚基类的目的是(　　)。

　　A. 简化程序　　　　　　　　　　B. 消除二义性
　　C. 提高运行效率　　　　　　　　D. 减少目标代码

【答】 B

(12) 继承具有(　　)，即当基类本身也是某一个类的派生类时，底层的派生类会自动继承间接基类的成员。

　　A. 规律性　　　B. 传递性　　　C. 重复性　　　D. 多样性

【答】 B

(13) 在派生类构造函数的初始化列表中不能包含(　　)。

　　A. 基类的构造函数　　　　　　　B. 基类对象成员的初始化
　　C. 派生类对象成员的初始化　　　D. 派生类中新增数据成员的初始化

【答】 D

(14) 在公有派生情况下，下面有关派生类对象和基类对象的关系不正确的是(　　)。

　　A. 派生类的对象可以赋给基类的对象
　　B. 派生类的对象可以初始化基类的引用
　　C. 派生类的对象可以直接访问基类中的成员
　　D. 派生类的对象的地址可以赋给指向基类的指针

【答】 C

(15) 有以下类定义：

```
class MyBASE{
        int k;
public:
    void set(int n){ k = n;}
    int get()const{ return k; }
};
class MyDERIVED: protected MyBASE{
protected:
    int j;
public:
    void set(int m, int n){ MyBASE::set(m); j = n;}
    int get()const{ return MyBASE::get() + j; }
};
```

则类 MyDERIVED 中 protected 访问权限的数据成员和成员函数的个数是(　　)。

　　A. 4　　　　　B. 3　　　　　C. 2　　　　　D. 1

【答】 B

(16) 有以下程序：

```
#include <iostream>
using namespace std;
class A {
public:
    A() { cout << "A"; }
};
class B { public: B() { cout << "B"; } };
class C : public A {
    B b;
public:
    C() { cout << "C"; }
};
int main() {  C obj;   return 0; }
```

执行后的输出结果是（   ）。

  A．CBA   B．BAC   C．ACB   D．ABC

【答】 D

(17) 有以下类定义：

```
class XA{
    int x;
public:
    XA(int n){ x = n; }
};
class XB: public XA{
    int y;
public:
    XB(int a, int b);
};
```

在构造函数 XB 的下列定义中，正确的是（   ）。

  A．XB::XB(int a,int b): x(a), y(b){ }

  B．XB::XB(int a,int b): XA(a), y(b){ }

  C．XB::XB(int a,int b): x(a), XB(b){ }

  D．XB::XB(int a,int b): XA(a), XB(b){ }

【答】 B

(18) 有以下程序：

```
#include <iostream>
using namespace std;
class BASE{
    public:
        ~BASE(){ cout <<"BASE"; }
};
class DERIVED: public BASE {
    public:
        ~DERIVED(){ cout <<"DERIVED"; }
};
int main(){DERIVED x; return 0 ;}
```

执行后的输出结果是（　　）。

　　A. BASE　　　　　　　　　　B. DERIVED
　　C. BASEDERIVED　　　　　　D. DERIVEDBASE

【答】 D

(19) 有以下程序：

```
#include <iostream>
using namespace std;
class Base {
public:
    void fun(){cout<<"Base::fun"<< endl;}
};
class Derived:public Base {
public:
    void fun() {
        _____
        cout<<"Derived::fun"<< endl;
    }
};
int main()  {
    Derived d;
    d.fun();
    return 0;
}
```

已知其执行后的输出结果为：

```
Base::fun
Derived::fun
```

则在程序中横线处应填入的语句是（　　）。

　　A. Base.fun();　　B. Base::fun();　　C. Base->fun();　　D. fun();

【答】 B

(20) 有以下程序：

```
#include <iostream>
using namespace std;
class A{
public:
    A(){cout<<"A";}
    ~A(){cout<<"-A";}
};
class B:public A{
    A * p;
public:
    B(){cout<<"B"; p = new A();}
    ~B(){cout<<"-B";  delete  p;}
};
int main(){
    B obj;
    return 0;
}
```

执行这个程序的输出结果是(　　)。

  A．BAA-A-B-A   B．ABA-B-A-A
  C．BAA-B-A-A   D．ABA-A-B-A

【答】 B

(21) 有以下程序：

```cpp
#include <iostream>
using namespace std;
class Base
{
private:
    void fun1()const{ cout<<"fun1";}
protected:
    void fun2()const{ cout<<"fun2";}
public:
    void fun3()const{ cout<<"fun3";}
};
class Derived : protected Base
{
public:
    void fun4()const{ cout<<"fun4";}
};
int main()
{
    Derived obj;
    obj.fun1();              //①
    obj.fun2();              //②
    obj.fun3();              //③
    obj.fun4();              //④
    return 0;
}
```

其中有语法错误的语句是(　　)。

  A．①②③④   B．①②③   C．②③④   D．①④

【答】 B

(22) 有以下类定义：

```cpp
class MyBase
{
private:
    int k;
public:
    MyBase(int n = 0):k(n){}
    int value()const{ return k; }
};

class MyDerived:MyBase{int j;
public:
    MyDerived(int i):j(i){}
    int getK()const{ return k;}
    int getj()const{return j;}
};
```

编译时发现有一处语法错误,对这个错误最准确的描述是( )。

A. 函数 getK 试图访问基类的私有成员变量 K

B. 在类 MyDerived 的定义中,基类名 MyBase 前缺少关键字 public、protected 或 private

C. 类 MyDerived 缺少一个无参的构造函数

D. 类 MyDerived 构造的函数没有对基数数据成员 K 进行初始化

【答】 A

(23) 有以下程序:

```cpp
#include<iostream>
using namespace std;
class Base{
protected:
    Base(){cout<<'A';}
    Base(char c){cout<<c;}
};

class Derived:public Base{
public:
    Derived(char c){cout<<c;}
};
int main()
{
    Derived d1('B');
    return 0;
}
```

执行这个程序屏幕上将显示( )。

A. B　　　　B. BA　　　　C. AB　　　　D. BB

【答】 C

3. 简答题

(1) 派生类如何实现对基类私有成员的访问?

【答】 无论使用任何一种继承方式,基类的私有成员都不允许外部函数直接访问,也不允许派生类的成员函数直接访问,但是可以通过基类的公有成员函数间接访问该类的私有成员。

(2) 什么是赋值兼容?它会带来什么问题?

【答】 赋值兼容是指在公有派生的情况下,一个派生类对象可以作为基类的对象来使用。赋值兼容又称为类型赋值兼容或类型适应。

在 C++ 中,赋值兼容主要指以下 3 种情况:

① 派生类对象可以赋值给基类对象。

② 派生类对象可以初始化基类的引用。

③ 派生类对象的地址可以赋给指向基类的指针。

由于派生类对象中包含基类子对象,所以这种引用方式是安全的,但是这种方法只能引用从基类继承的成员。如果试图通过基类指针引用那些只有在派生类中才有的成员,编译器将会报告语法错误。

(3) 多重继承时,构造函数和析构函数的执行是如何实现的?

【答】 多重继承时,构造函数的执行顺序是先执行基类的构造函数,再执行对象成员的构造函数,最后执行派生类的构造函数。

在多个基类之间则严格按照派生类声明时从左到右的顺序执行各基类的构造函数,而析构函数的执行顺序正好与构造函数的执行顺序相反。

(4) 继承与组合之间有什么区别和联系?

【答】 继承描述的是一般类与特殊类的关系,类与类之间体现的是"is a kind of",即如果在逻辑上 A 是 B 的一种,允许 A 继承 B 的功能和属性。例如汽车(automobile)是交通工具(vehicle)的一种,小汽车(car)是汽车的一种,那么类 automobile 可以从类 vehicle 派生,类 car 可以从类 automobile 派生。

组合描述的是整体与部分的关系,类与类之间体现的是"is a part of",即如果在逻辑上 A 是 B 的一部分,则允许 A 和其他数据成员组合为 B。例如,发动机、车轮、电池、车门、方向盘、底盘都是小汽车的一部分,它们组合成汽车,而不能说汽车是发动机的一种。

在 C++ 中,类的继承与类的组合很相似,继承和组合既有区别,又有联系,主要表现在描述的关系不同。某些比较复杂的类既需要使用继承,也需要使用组合,二者一起使用。

在某些情况下,继承与组合的实现还可以互换。在多继承时,一个派生类有多个直接基类,派生类实际上是所有基类属性和行为的组合。派生类是对基类的扩充,派生类成员一部分是从基类中而来,因此派生类组合了基类。既然这样,派生类也可以通过组合类实现。什么时候使用继承,什么时候使用组合,要根据问题中类与类之间的具体关系顺其自然,权衡考虑。

(5) 什么是虚基类?它有什么作用?

【答】 在多继承中,当派生类的部分或全部直接基类又是从另一个共同基类派生而来时,这些直接基类中从上一级共同基类继承来的成员就拥有相同的名称。在派生类的对象中,同名数据成员在内存中同时拥有多个副本,同一个成员函数会有多个映射,出现二义性,因此,C++将共同基类设置为虚基类。虚基类使得从不同的路径继承过来的同名数据成员在内存中只有一个副本,同一个函数名也只有一个映射。这样不仅解决了二义性问题,也节省了内存,避免了数据不一致的问题。

4. 程序分析题(写出运行结果)

(1)

```cpp
#include <iostream>
using namespace std;
class B1{
public:
    B1(int i)
    {cout <<"constructing B1  " << i << endl;}
    ~B1()
    {cout <<"destructing B1   "<< endl;}
};
class B2 {
public:
    B2(int i)
    {cout <<"constructing B2  " << i << endl;}
    ~B2()
    {cout <<"destructing B2   "<< endl;}
```

```
};
class B3 {
public:
    B3()
    {cout <<"constructing B3    * "<< endl;}
    ~B3()
    {cout <<"destructing B3"<< endl;}
};
class C:public B2,public B1,public B3 {
public:
    C(int a,int b,int c,int d):B1(a),memberB2(d),memberB1(c),B2(b){}
private:
    B1 memberB1;
    B2 memberB2;
    B3 memberB3;
};
int main()
{
    C obj(1,2,3,4);
    return 0;
}
```

【答】 运行结果：

```
constructing B2   2
constructing B1   1
constructing B3    *
constructing B1   3
constructing B2   4
constructing B3    *
destructing B3
destructing B2
destructing B1
destructing B3
destructing B1
destructing B2
```

（2）

```
#include <iostream>
using namespace std;
class A{
    public: A(){a = 0;}
            A(int i){a = i;}
            void Print(){cout << a <<",";}
            int Geta(){return a;}
    private:int a;
};
class B:public A{
    public:B(){b = 0;}
           B(int i,int j,int k);
           void Print();
```

```
        private:int b;
              A aa;
};
B::B(int i,int j,int k):A(i),aa(j){b = k;}
void B::Print(){
    A::Print();
    cout << b <<","<< aa.Geta()<< endl;
}
int main(){
    B bb[2];
    bb[0] = B(1,2,5);
    bb[1] = B(3,4,7);
    for(int i = 0;i < 2;i++)
        bb[i].Print();
    return 0;
}
```

【答】 运行结果：

```
1,5,2
3,7,4
```

5. 程序填空题

在下面横线处填上合适的内容，完成类 B 的定义。

```
# include < iostream >
using namespace std;
class A{
    public:A(){a = 0;}
          A(int i){a = i;}
          void print(){cout << a <<",";}
    private:int a;
};
class B:public A{
    public:B(){b1 = b2 = 0}
        B(    ①    ){b1 = i;b2 = 0;}
        B(int i,int j,int k):    ②    {b1 = j;b2 = k;}    //使 a 的值为 i
        void print(){A::print();cout << b1 <<","<< b2 << endl;}
        private:int b1,b2;
}
```

【答】 ① int i
      ② A(i)

6. 改错题

(1) 下面程序中有一处错误，请用横线标出错误所在行并说明错误原因，使程序的输出结果为：

```
m = 0
m = 10
```

```
# include < iostream >
```

```
using namespace std;
class c0 {
    int m;
public:
    void print(){ cout <<"m = "<< m << endl; }
};
class c1:public c0 {
public:
    c1(int t)
    {m = t;}
};
int main() {
    c1 obj1(0);
    obj1.print();
    c1 obj2(10);
    obj2.print();
    return 0;
}
```

【答】 "m＝t;"错误,m 是基类 c0 的私有数据成员,派生类的构造函数不能访问基类的私有成员。

改正:将 m 声明为类 c0 的公有数据成员。

```
public:
    int m;
```

(2) 指出下面程序段中的错误,并说明出错原因。

```
class one{
    private:
        int a;
    public:
        void func(two&);
};
class two{
    private:
        int b;
        friend void one::func(two&);
};
void one::func(two& r)
{
    a = r.b;
}
```

【答】 "void func(two&);"错误,对对象的引用必须声明在前,引用在后。类 one 的公有成员函数 func()的形参是类 two 的引用对象,但类 two 的声明却在类 one 之后。

改正:在类 one 之前加前向引用说明"class two;"。

(3) 指出下面程序段中的错误,并说明出错原因。

```
class A{
    public:   void fun(){cout <<"a.fun"<< endl;}
};
```

```
class B{
    public: void fun(){cout <<"b.fun"<< endl;}
            void gun(){cout <<"b.gun"<< endl;}
};
class C:public A,public B{
    private:int b;
    public:void gun(){cout <<"c.gun"<< endl;}
           void hun(){fun();}
};
```

【答】 "void hun(){fun();}"错误,出现二义性,用基类名和类作用域分辨符加以指定,消除二义性。

改正:void hun(){A::fun();}或者 void hun(){B::fun();}

(4)下面程序中有一处错误,请用横线标出错误所在行并说明错误原因。

```
#include <iostream>
using namespace std;
class A{
    public:A(const char * nm){strcpy(name,nm);}
    private:char name[80];
};
class B:public A{
    public:B(const char * nm):A(nm){ }
            void PrintName()const;
};
void B::PrintName()const{
    cout <<"name:"<< name << endl;
}
int main(){
    B b1("wang li");
    b1.PrintName();
    return 0;
}
```

【答】 "cout <<"name:"<< name << endl;"错误,不能直接访问基类的私有数据成员。

改正:public: char name[80];

或通过增加一个基类的公有函数成员,通过该成员函数间接访问基类的私有数据成员。

7. 编程题

(1)定义一个 Point 类,派生出 Rectangle 类和 Circle 类,计算各派生类对象的面积 Area()。

【答】 程序如下:

```
#include <iostream>
using namespace std;
class Point                    //定义基类 Point
{
public:
    Point(float xx,float yy)
    {
        x = xx;
        y = yy;
```

```cpp
    }
private:
    float x;
    float y;
};
class Rectangle :public Point            //定义派生类 Rectangle
{
public:
    Rectangle(float xx,float yy,float w,float h);
    float Area()
    {
        return width * high;
    }
private:
    int width;
    int high;
};
Rectangle::Rectangle(float xx,float yy,float w, float h):Point(xx,yy)
{
    width = w;
    high = h;
}
class Circle :public Point               //定义派生类 Circle
{
public:
    Circle(float xx,float yy,float r);
    float Area()
    {
        return 3.14 * radius * radius;
    }
private:
    float radius;
};
Circle::Circle(float xx,float yy,float r):Point(xx,yy) { radius = r; }

int main()
{
    Rectangle R(1,2,3,4);
    cout <<"R. Area() = "<< R. Area()<< endl;
    Circle C(5,6,7);
    cout <<"C. Area() = "<< C. Area()<< endl;
    return 0;
}
```

运行结果：

```
R. Area() = 12
C. Area() = 153.86
```

(2) 设计一个建筑物类 Building，由它派生出教学楼类 TeachBuilding 和宿舍楼类

DormBuilding，前者包括教学楼编号、层数、教室数、总面积等基本信息，后者包括宿舍楼编号、层数、房间数、总面积和容纳学生总人数等基本信息。

【答】 程序如下：

```cpp
#include <iostream>
using namespace std;

class Building                          //定义基类 Building
{
public:
    Building(char *name, int floor, int room, int area)
    {
        strcpy(this->name, name);
        this->floor = floor;
        this->room = room;
        this->area = area;
    }
    void display()
    {
        cout<<"建筑物名称："<<name<<endl;
        cout<<"层数："<<floor<<endl;
        cout<<"教室(房间)数："<<room<<endl;
        cout<<"面积："<<area<<endl;
    }
protected:
    char name[20];              //名称
    int floor;                  //层数
    int room;                   //房间数
    float area;                 //面积
};

class TeachBuilding:public Building    //定义派生类 TeachBuilding
{
public:
    TeachBuilding(char *name, int floor, int room, int area, char *function):Building(name, floor, room, area)
    {
        strcpy(this->function, function);
    }
    void display()
    {
        Building::display();
        cout<<"建筑物功能："<<function<<endl;
    }
private:
    char function[20];          //建筑物的功能
};

class DormBuilding:public Building    //定义派生类 DormBuilding
{
public:
```

```
        DormBuilding(char * name,int floor,int room,int area,int peoples)
:Building(name,floor,room,area)
        {
            this->peoples = peoples;
        }
        void display()
        {
            Building::display();
            cout<<"容纳人数: "<<peoples<<endl;

        }
private:
        int peoples;                           //容纳人数
};

int main()
{
    TeachBuilding TB("5号楼",5,40,4000,"物理教学楼");
    TB.display();
    cout<<endl;
    DormBuilding DB("宿舍16栋",7,140,1200,560);
    DB.display();
    return 0;
}
```

运行结果：

```
建筑物名称: 5号楼
层数: 5
教室(房间)数: 40
面积: 4000
建筑物功能: 物理教学楼

建筑物名称: 宿舍16栋
层数: 7
教室(房间)数: 140
面积: 1200
容纳人数: 560
```

(3) 定义并描述一个 Table 类和一个 Circle 类，由它们共同派生出 RoundTable 类。

【答】 程序如下：

```
#include<iostream>
using namespace std;
class Circle                                   //定义基类 Circle
{
public:
    Circle(float r) { radius = r; }
    float GetArea() { return 3.14*radius*radius; }
private:
```

```
    float radius;
};
class Table                              //定义基类 Table
{
public:
    Table(float h) { high = h; }
    float GetHigh() { return high; }
private:
    int high;
};

class RoundTable :public Circle, public Table      //定义派生类 RoundTable
{
public:
    RoundTable(float r,float h,char * c):Circle(r),Table(h)
    {
        strcpy(color,c);
    }
    char * GetColor() { return color; }
private:
    char color[20];
};
int main()
{
    RoundTable RT(3,5,"橘黄");
    cout <<"高度："<< RT.GetHigh()<< endl;
    cout <<"面积："<< RT.GetArea()<< endl;
    cout <<"颜色："<< RT.GetColor()<< endl;
    return 0;
}
```

运行结果：

```
高度：5
面积：28.26
颜色：橘黄
```

（4）定义并描述一个人员类 Person，它派生出学生类 Student 和教师类 Teacher，学生类和教师类又共同派生出在职读书的教师类 Stu_Teach。人员类有姓名、性别、身份证号、出生年月等信息；学生类有专业等信息；教师类有职称等信息。

【答】 程序如下：

```
#include <iostream>
#include <string>
using namespace std;

class Person                             //定义虚基类 Person
{
public:
```

```cpp
        Person(char * name, char sex, char * ID, char * birthday)
        {
            strcpy(this -> name, name);
            this -> sex = sex;
            strcpy(this -> ID, ID);
            strcpy(this -> birthday, birthday);
        }
        void display()
        {
            cout <<"name:"<< name << endl;
            cout <<"sex:"<< sex << endl;
            cout <<"ID:"<< ID << endl;
            cout <<"birthday:"<< birthday << endl;
        }
    private:
        char name[20];              //姓名
        char sex;                   //性别
        char ID[20];                //身份证号
        char birthday[12];          //出生年月
};

class Teacher :virtual public Person      //定义派生类 Teacher
{
public:
        Teacher(char * name, char sex, char * ID, char * birthday, char * position)
        :Person(name, sex, ID, birthday)
        {
            strcpy(this -> position, position);
        }
        void display()
        { cout <<"position:"<< position << endl; }
private:
        char position[12];          //职称
};

class Student :virtual public Person      //定义派生类 Student
{
public:
        Student(char * name, char sex, char * ID, char * birthday, char * major)
        :Person(name, sex, ID, birthday)
        {
            strcpy(this -> major, major);
        }
        void display(){ cout <<"major:"<< major << endl; }
private:
        char major[20];             //专业
};

class Stu_Teach:public Teacher, public Student
{
```

```cpp
public:
    Stu_Teach(char * name,char sex,char * ID,char * birthday,char * position, char * major)
        :Person(name,sex, ID, birthday),Student(name,sex,ID,birthday, major),Teacher(name,sex,ID,birthday,position)
    { }
    void display()
    {
        Person::display();
        Student::display();
        Teacher::display();
    }
};

int main()
{
    Stu_Teach ST("Liu_xiaopeng",'M',"428120198005272487","1976-05-27",
        "Professor", "Computer Science");
    ST.display();
    return 0;
}
```

运行结果：

```
name:Liu_xiaopeng
sex:M
ID:428120198005272487
birthday:1976-05-27
major:Computer Science
position:Professor
```

(5) 利用 Clock 类定义一个带"AM""PM"的新时钟类 NewClock。

【答】 程序如下：

```cpp
#include <iostream>
using namespace std;
class Clock                              //定义时钟类 Clock
{
public:
    Clock(int h,int m,int s)
    {
        hour = (h>23?0:h);
        minute = (m>59?0:m);
        second = (s>59?0:s);
    }
    void run()                           //时钟运行
    {
        second = second + 1;
        if(second>59)
        {
```

```cpp
                second = 0;
                minute = minute + 1;
            }
            if(minute > 59)
            {
                minute = 0;
                hour = hour + 1;
            }
            if(hour > 23)
                hour = 0;
        }
        void ShowTime()                    //显示时间
        {
            cout <<"Now: "<< hour <<":"<< minute <<":"<< second << endl;
        }
        int gethour() {    return hour; }
        int getminute() {    return minute; }
        int getsecond() {    return second; }
    private:
        int hour;                          //时
        int minute;                        //分
        int second;                        //秒
};

class NewClock :public Clock               //定义新的 NewClock 类
{
public:
    NewClock(int h, int m, int s):Clock(h, m, s)
    {
        if(h >= 12&&h < 24)
            AP = 'P';
        else
            AP = 'A';
    }
    void run() {
        Clock::run();
        Clock::gethour()>= 12?AP = 'P':AP = 'A';
    }
    void ShowTime()                        //重新显示时间
    {
cout << Clock::gethour() % 12 <<":"<< Clock::getminute()<<":"<< Clock::
        getsecond();
        cout << AP <<"M"<< endl;
    }
private:
    char AP;                               //上午/下午
};
int main()
{
    NewClock C1(10,30,0);
    C1.ShowTime();
    for( int i = 0; i < 6000; i++)
```

```
        C1.run();
        C1.ShowTime();
        return 0;
}
```

运行结果：

```
10:30:0AM
0:10:0PM
```

(6) 利用 Clock 类和 Date 类定义一个带日期的时钟类 ClockWithDate，且对该类对象能进行增加秒数的操作。

【答】 程序如下：

```
#include <iostream>
using namespace std;
class Clock                              //定义时钟类 Clock
{
    public:
        Clock(int h, int m, int s);
        void run();
        void ShowTime()                  //显示时间
        {
            cout<<"Now: "<< hour <<":"<< minute <<":"<< second << endl;
        }
        int gethour()    {  return hour;   }
        int getminute()  {  return minute; }
        int getsecond()  {  return second; }
    private:
        int hour;                        //时
        int minute;                      //分
        int second;                      //秒
};

Clock::Clock(int h, int m, int s)    {
    hour = (h > 23?0:h);
    minute = (m > 59?0:m);
    second = (s > 59?0:s);
}

void Clock::run()                        //时钟运行
{
    second = second + 1;
    if(second > 59) {
        second = 0;
        minute = minute + 1;
    }
    if(minute > 59) {
        minute = 0;
```

```cpp
            hour = hour + 1;
        }
        if( hour > 23 )
            hour = 0;
    }

class Date                                    //定义日期类 Date
{
public:
    Date( int y = 1996, int m = 1, int d = 1 );
    int days( int year, int month ) ;
    void NewDay();
    void display() {
        cout << year <<" - "<< month <<" - "<< day << endl;
    }
private:
    int year;                                 //年
    int month;                                //月
    int day;                                  //日
};

int Date::days( int year, int month ) {
    bool leap;
    if( ( year % 400 == 0 ) || ( year % 4 == 0 && year % 100 != 0 ) )
        leap = true;
    else
        leap = false;
    switch( month )
    {
        case 1:
        case 3:
        case 5:
        case 7:
        case 8:
        case 10:
        case 12:
            return 31;
        case 4:
        case 6:
        case 9:
        case 11:
            return 30;
        case 2:
            if( leap )                        //是否闰月
                return 29;
            else
                return 28;
            break;
    }
}

Date::Date( int y, int m, int d ){
```

```cpp
        if(m > 12 || m < 1){
            cout <<"Invalid month! "<< endl;
            m = 1;
        }
        if(d > days(y,m)) {
            cout <<"Invalid day! "<< endl;
            d = 1;
        }
        day = d;
        year = y;
        month = m;
}

void Date::NewDay(){
    if (day < days(year,month))
        day++;
    else {
        day = 1;
        month++;
        if(month == 13) {
            year++;
            month = 1;
        }
    }
}

class ClockWithDate :public Clock,public Date    //定义 ClockWithDate 类
{
  public:
        ClockWithDate(int hour, int minute, int second, int year, int month, int day): Clock(hour,
minute,second),Date(year,month,day) {}
        void run();
        void ShowTime()                    //显示时间和日期
        {
            Clock::ShowTime();
            display();
        }
};

void ClockWithDate::run(){
    Clock::run();
    if(gethour() == 0&&getminute() == 0&&getsecond() == 0)
        NewDay();                          //新的一天
}

int main(){
    ClockWithDate CD(7,59,59,2009,12,31);
    CD.ShowTime();
    for(int i = 0;i < 3600 * 24 + 1000;i++)
        CD.run();
```

```
        cout <<"after 3600 * 24 + 1000 seconds: ";
        CD.ShowTime();
        return 0;
}
```

运行结果：

```
Now: 7:59:59
2009 - 12 - 31
after 3600 * 24 + 1000 seconds: Now: 7:16:39
2010 - 1 - 1
```

(7) 扩充 String 类，增加对字符串编辑的函数，例如替换某个字符、替换某个字符串、删除某个字符、删除某个字符串等。

【答】 程序如下：

```
#include <iostream>
using namespace std;
class String
{
protected:
    char * mystr;                    //字符串
    int len;                         //字符串的长度
public:
    String()
    {
        len = 0;
        mystr = NULL;
    }
    String (const char * p)
    {
        len = strlen (p);
        mystr = new char [len + 1];
        strcpy (mystr,p);
    }
    String (String & r)              //定义拷贝构造函数,防止浅拷贝
    {
        len = r.len;
        if (len!= 0)
        {
            mystr = new char[len + 1];
            strcpy (mystr,r.mystr);
        }
    }
    ~String()
    {
        if (mystr!= NULL)
        {
            delete []mystr;
```

```cpp
            mystr = NULL;
            len = 0;
        }
    }
    char * GetStr()const {   return mystr;   }
    void ShowStr()const { cout << mystr;}
    bool IsSubstring(const char * str);              //判断 str 是否为当前串的子串
    bool IsSubstring(const String &str);             //判断 String 类对象 str 是否为当前串的子串
    int str2num();                                   //把字符串转换成数
    void toUppercase();                              //把字符串转换成大写字母
};
bool String::IsSubstring(const char * str)           //判断 str 是否为当前串的子串
{
    int i,j;
    for (i = 0;i < len;i++)
    {
        int k = i;
        for (j = 0;str[j];j++,k++)
            if (str[j]!= mystr[k]) break;
        if (!str[j])   return true;
    }
    return false;
}
bool String::IsSubstring(const String &str)
{
    char * pstr = str.mystr;
    int plen = str.len;
    for (int i = 0;i < len;i++)
    {
        int k = i;
        for (int j = 0;j < plen;j++,k++)
            if (pstr[j]!= mystr[k])   break;
        if (j == plen)   return true;
    }
    return false;
}

class EditString: public String
{
public:
    EditString():String(){}
    EditString(const char * p):String(p){}
    EditString(String &r):String(r){}
    int IsSubstring(int start,const char * str);
    void EditChar(char s,char d);                    //用字符 d 代替所有的字符 s
    void EditSub(char * subs,char * subd);           //用字符串 subd 代替所有的字符串 subs
    void DeleChar(char ch);                          //将所有的字符 ch 删除
    void DeleSub(char * sub);                        //将所有的字符串 sub 删除
};
void EditString::EditChar(char s,char d)
{
```

```cpp
    for (int i = 0;i < len;i++)
        if (mystr[i] == s) mystr[i] = d;
}
void EditString::DeleChar(char ch)        //将所有的字符 ch 删除
{
    for (int i = 0,j = 0;j < len;j++)
        if ( mystr[j]!= ch)
        {
            mystr[i] = mystr[j];
            i++;
        }
    mystr[i] = '\0';
}

//从 start 处开始判断子串函数,是则返回第一次所在的下标位置,否则返回 -1
int EditString::IsSubstring(int start,const char * str)
{
    for (int i = start;i < len;i++)
    {
        int k = i;
        for (int j = 0;str[j]&&k < len;j++,k++)
            if (str[j]!= mystr[k]) break;
        if (!str[j])   return i;
    }
    return -1;
}

void EditString::EditSub(char * subs, char * subd)
{ //为避免数组元素移动,使用一个较大的临时数组
    char * temp = new char[len + len + len];
    int i = 0;                    //指向 mystr
    int el = 0;                   //指向 temp
    while (i < len)
    {
        int pos = IsSubstring(i,subs);
        if (pos == -1)
        {
            for (int j = i;j < len;j++)
                temp[el++] = mystr[j];
            break;
        }
        for (int j = i;j < pos;j++)
            temp[el++] = mystr[j];
        for (j = 0;subd[j];j++)
            temp[el++] = subd[j];
        i = pos + strlen(subs);
    }
    temp[el] = '\0';
    len = strlen(temp);
    if(mystr!= NULL)
        delete[] mystr;
    if (len > 0){
        mystr = new char[len + 1];
        strcpy(mystr,temp);
```

```cpp
        }
        delete[]temp;
}

void EditString::DeleSub(char * sub)            //将所有的字符串 sub 删除
{
    if (!(String::IsSubstring(sub)))   return ; //调用父类判断函数
    char * temp = new char[len + 1];
    int i = 0;                                  //指向 mystr
    int el = 0;                                 //指向 temp
    while (i < len)
    {
        int pos = IsSubstring(i,sub);
        if (pos == -1)
        {
            for (int j = i;j < len;j++)
                temp[el++] = mystr[j];
            break;
        }
        for (int j = i;j < pos;j++)
            temp[el++] = mystr[j];
        i = pos + strlen(sub);
    }
    temp[el] = '\0';
    len = el;
    strcpy(mystr,temp);
    delete[]temp;
}

int main()
{
    EditString s1("abcdddddda");
    cout <<"s1: ";
    s1.ShowStr();

    cout <<"\nAfter call function EditChar('d','r'): ";
    s1.EditChar('d','r');                       //测试字符替代函数
    s1.ShowStr();

    cout <<"\nAfter call function DeleChar('r'): ";
    s1.DeleChar('r');                           //测试字符删除函数
    s1.ShowStr();

    cout <<"\nAfter call function EditSub(\"bca\",\"12345\"): ";
    s1.EditSub("bca","12345");                  //测试字符串代替函数
    s1.ShowStr();

    cout <<"\nAfter call function DeleSub(\"234\"): ";
    s1.DeleSub("234");                          //测试字符串删除函数
    s1.ShowStr();
    return 0;
}
```

运行结果:

```
s1: abcddddda
After call function EditChar('d','r'): abcrrrrra
After call function DeleChar('r'): abca
After call function EditSub("bca","12345"): a12345
After call function DeleSub("234"): a15
```

（8）编写一个程序,实现小型公司的工资管理。该公司主要有 4 类人员,即经理（manager）、技术人员（technician）、推销员（salesman）、销售经理（salesmanager）。这些人员都是职员（employee）,有编号、姓名、月工资信息。月工资的计算方法是经理固定月薪 8000 元；技术人员每小时 100 元；推销员按当月销售额的 4% 提成；销售经理既拿固定月工资 5000 元,又拿销售提成,销售提成为所管辖部门当月销售额的 5‰。要求编程计算月工资并显示全部信息。

【答】 程序如下：

```cpp
#include<iostream>
using namespace std;
class employee
{
protected:
    char name[20];                          //姓名
    int individualEmpNo;                    //个人编号
    double accumPay;                        //月薪总额
    static int employeeNo;                  //本公司职员编号目前的最大值
public:
    employee();                             //构造函数
    void pay();                             //计算月薪函数
    void SetName(char *);                   //设置姓名函数
    char * GetName();                       //提取姓名函数
    int GetindividualEmpNo();               //提取编号函数
    double GetaccumPay();                   //提取月薪函数
};

class technician:public employee            //兼职技术人员类
{
private:
    double hourlyRate;                      //每小时酬金
    int workHours;                          //当月工作时数
public:
    technician();                           //构造函数
    void SetworkHours(int wh);              //设置工作时数
    void pay();                             //计算月薪函数
};

class salesman:virtual public employee      //兼职推销员类
{
protected:
    double CommRate;                        //按销售额提取酬金的百分比
```

```cpp
        double sales;                                    //当月销售额
    public:
        salesman();                                      //构造函数
        void Setsales(double sl);                        //设置销售额
        void pay();                                      //计算月薪函数
};

class manager:virtual public employee                    //经理类
{
protected:
        double monthlyPay;                               //固定月薪数
public:
        manager();                                       //构造函数
        void pay();                                      //计算月薪函数
};

class salesmanager:public manager,public salesman        //销售经理类
{
public:
        salesmanager();                                  //构造函数
        void pay();                                      //计算月薪函数
};

int employee::employeeNo = 1000;                         //员工编号基数为1000
employee::employee()
{                                                        //新输入的员工编号为目前最大编号加1
    individualEmpNo = employeeNo++;
    accumPay = 0.0;                                      //月薪总额初值为0
}

//employee::~employee(){}
void employee::pay(){}                                   //计算月薪,空函数
void employee::SetName(char * names) { strcpy(name,names); }   //设置姓名
char * employee::GetName(){return name;}                 //获取姓名

int employee::GetindividualEmpNo(){return individualEmpNo;}  //获取成员编号

double employee::GetaccumPay(){    return accumPay; }    //获取月薪

technician::technician()
{    hourlyRate = 100;}                                  //每小时酬金100元

void technician::SetworkHours(int wh)
{    workHours = wh;}                                    //设置工作时间

void technician::pay()
{    accumPay = hourlyRate * workHours;}                 //计算月薪,按小时计酬

salesman::salesman()
{    CommRate = 0.04;}                                   //销售提成比例4%

void salesman::Setsales(double sl)
```

```cpp
{     sales = s1;}                                      //设置销售额

void salesman::pay()
{     accumPay = sales * CommRate;}                     //月薪 = 销售提成

manager::manager()
{     monthlyPay = 8000;}                               //固定月薪 8000 元

void manager::pay()
{ accumPay = monthlyPay; }                              //月薪总额即固定月薪数

salesmanager::salesmanager()
{     monthlyPay = 5000;
     CommRate = 0.005;
}

void salesmanager::pay()
{     //月薪 = 固定月薪 + 销售提成
     accumPay = monthlyPay + CommRate * sales;
}

int main()
{
     manager m1;
     technician t1;
     salesmanager sm1;
     salesman s1;
     m1.SetName("Li Huanying");                         //设置雇员 m1 的姓名
     t1.SetName("Zhang Xiaofei");                       //设置雇员 t1 的姓名
     sm1.SetName("Shen Teng");                          //设置雇员 sm1 的姓名
     s1.SetName("Jia Ling");                            //设置雇员 s1 的姓名

     m1.pay();                                          //计算 m1 月薪

     cout <<"请输入兼职技术人员："<< t1.GetName()<<"本月的工作时数：";
     int wt;
     cin >> wt;                                         //输入 t1 本月的工作时数
     t1.SetworkHours(wt);                               //设置 t1 本月的工作时数
     t1.pay();                                          //计算 t1 的月薪

     cout <<"请输入销售经理："<< sm1.GetName()<<"所管辖部门本月的销售总额：";
     float s;
     cin >> s;                                          //输入 sm1 所管辖部门本月的销售总额
     sm1.Setsales(s);                                   //设置 sm1 所管辖部门本月的销售总额
     sm1.pay();                                         //计算 sm1 的月薪

     cout <<"请输入推销员 "<< s1.GetName()<<"本月的销售额：";
     cin >> s;                                          //输入 s1 本月的销售额
     s1.Setsales(s);                                    //设置 s1 本月的销售额
     s1.pay();                                          //计算 s1 的月薪

     //显示 m1 的信息
```

```
        cout <<"编号: "<< m1.GetindividualEmpNo()<< m1.GetName()
            <<"本月工资: "<< m1.GetaccumPay()<< endl;
    //显示 t1 的信息
        cout <<"编号: "<< t1.GetindividualEmpNo()<< t1.GetName()
            <<"本月工资: "<< t1.GetaccumPay()<< endl;
    //显示 sm1 的信息
        cout <<"编号: "<< sm1.GetindividualEmpNo()<< sm1.GetName()
            <<"本月工资: "<< sm1.GetaccumPay()<< endl;
    //显示 s1 的信息
        cout <<"编号: "<< s1.GetindividualEmpNo()<< s1.GetName()
            <<"本月工资: "<< s1.GetaccumPay()<< endl;
    return 0;
}
```

运行结果：

```
请输入兼职技术人员: Zhang Xiaofei 本月的工作时数:320↙
请输入销售经理: Shen Teng 所管辖部门本月的销售总额:2400000↙
请输入推销员 Jia Ling 本月的销售额:58000↙
编号: 1000Li Huanying 本月工资:8000
编号: 1001Zhang Xiaofei 本月工资:32000
编号: 1002Shen Teng 本月工资:17000
编号: 1003Jia Ling 本月工资:23200
```

1.7 习题 7 解答

1. 填空题

(1) 将一个函数调用链接上相应函数体的代码,这一过程称为**联编(绑定)**。

(2) C++支持两种多态性,即**静态多态性**和**动态多态性**。

(3) 在编译时就确定的函数调用称为**静态联编**,它通过使用**重载函数**实现。

(4) 在运行时才确定的函数调用称为**动态联编**,它通过**继承和虚函数**实现。

(5) 虚函数的声明方法是在函数原型前加上关键字 **virtual**。

(6) C++的静态多态性是通过**重载函数**实现的。

(7) C++的动态多态性是通过**虚函数**实现的。

(8) 当通过**基类指针**使用虚函数时,C++会在与对象关联的派生类中正确地选择重定义的函数。

(9) 如果一个类包含一个或多个纯虚函数,则该类为**抽象类**。

(10) 若以非成员函数形式为类 Bounce 重载 ! 运算符,其操作结果为 bool 型数据,则该运算符重载函数的原型是 **friend bool operate ! (Bounce)**;。

2. 选择题

(1) 下列运算符中,(　　)运算符在 C++中不能重载。

  A. &&　　　　B. [ ]　　　　C. ::　　　　D. new

【答】C

(2) 下列关于运算符重载的描述中,( )是正确的。

A. 运算符重载可以改变运算数的个数

B. 运算符重载可以改变优先级

C. 运算符重载可以改变结合性

D. 运算符重载不能改变语法结构

【答】D

(3) 如果表达式++i*k中的"++"和"*"都是重载的友元运算符,则采用运算符函数调用格式,该表达式还可表示为( )。

A. operator*(i.operator++(),k)   B. operator*(operator++(i),k)

C. i.operator++().operator*(k)   D. k.operator*(operator++(i))

【答】B

(4) 在下列成对的表达式中,运算符"+"的意义不相同的一对是( )。

A. 5.0+2.0 和 5.0+2   B. 5.0+2.0 和 5+2.0

C. 5.0+2.0 和 5+2   D. 5+2.0 和 5.0+2

【答】C

(5) 下列关于运算符重载的叙述中,正确的是( )。

A. 通过运算符重载可以定义新的运算符

B. 有的运算符只能作为成员函数重载

C. 若重载运算符+,则相应的运算符函数名是+

D. 重载一个二元运算符时必须声明两个形参

【答】B

(6) 已知在一个类体中包含函数原型"VOLUME operator-(VOLUME)const;",下列关于这个函数的叙述错误的是( )。

A. 这是运算符-的重载运算符函数

B. 这个函数所重载的运算符是一个一元运算符

C. 这是一个成员函数

D. 这个函数不改变类的任何数据成员的值

【答】B

(7) 在表达式x+y*z中,+是作为成员函数重载的运算符,*是作为非成员函数重载的运算符。下列叙述中正确的是( )。

A. operator+有两个参数,operator*有两个参数

B. operator+有两个参数,operator*有一个参数

C. operator+有一个参数,operator*有两个参数

D. operator+有一个参数,operator*有一个参数

【答】C

(8) 在C++程序中,对象之间的相互通信通过( )。

A. 继承实现   B. 调用成员函数实现

C. 封装实现   D. 函数重载实现

【答】B

(9) 下面是重载为非成员函数的运算符函数原型,其中错误的是(   )。
    A. Franction operator ＋（Franction ,Franction）;
    B. Franction operator －（Franction）;
    C. Franction &operator ＝（Franction &,Franction）;
    D. Franction &operator ＋＝（Franction&, Franction）;
【答】 B

(10) 当一个类的某个函数被说明为 virtual 时,该函数在该类的所有派生类中(   )。
    A. 都是虚函数
    B. 只有被重新说明时才是虚函数
    C. 只有被重新说明为 virtual 时才是虚函数
    D. 都不是虚函数
【答】 A

(11) (   )是一个在基类中说明的虚函数,它在该基类中没有定义,但要求任何派生类都必须定义自己的版本。
    A. 虚析构函数   B. 虚构造函数
    C. 纯虚函数   D. 静态成员函数
【答】 C

(12) 以下基类中的成员函数,表示纯虚函数的是(   )。
    A. virtual void vf(int);   B. void vf(int)＝0;
    C. virtual void vf()＝0;   D. virtual void yf(int){}
【答】 C

(13) 如果一个类至少有一个纯虚函数,那么就称该类为(   )。
    A. 抽象类   B. 虚基类
    C. 派生类   D. 以上都不对
【答】 A

(14) 下列关于纯虚函数和抽象类的描述,(   )是错误的。
    A. 纯虚函数是一种特殊的虚函数,它没有具体的定义
    B. 抽象类是指具有纯虚函数的类
    C. 若一个基类中说明有纯虚函数,该基类的派生类一定不再是抽象类
    D. 抽象类只能作为基类使用,其纯虚函数的定义由派生类给出
【答】 C

(15) 在下列描述中,(   )是抽象类的特性。
    A. 可以说明虚函数   B. 可以进行构造函数重载
    C. 可以定义友元函数   D. 不能定义其对象
【答】 D

(16) 抽象类应含有(   )。
    A. 至多一个虚函数   B. 至少一个虚函数
    C. 至多一个纯虚函数   D. 至少一个纯虚函数
【答】 D

(17) 类 B 是类 A 的公有派生类,类 A 和类 B 中都定义了虚函数 func(),p 是一个指向类 A 对象的指针,则 p—>A::func()将(   )。

  A. 调用类 A 中的函数 func()

  B. 调用类 B 中的函数 func()

  C. 根据 p 所指的对象类型确定调用类 A 中或类 B 中的函数 func()

  D. 既调用类 A 中的函数,也调用类 B 中的函数

【答】 A

(18) 在 C++中,用于实现运行时多态性的是(   )。

  A. 内联函数　　　B. 重载函数　　　C. 模板函数　　　D. 虚函数

【答】 D

(19) 对于类定义:

```
class A{
    public:
        virtual void func1(){}
        void func2(){}
};
class B:public A{
    public:
        void func1(){cout <<"class B func1"<< endl;}
        virtual void func2(){cout <<"class B func2"<< endl;}
};
```

下面叙述正确的是(   )。

  A. A::func2()和 B::func1()都是虚函数

  B. A::func2()和 B::func1()都不是虚函数

  C. B::func1()是虚函数,而 A::func2()不是虚函数

  D. B::func1()不是虚函数,而 A::func2()是虚函数

【答】 C

(20) 下列关于虚函数的说明中,正确的是(   )。

  A. 从虚基类继承的函数都是虚函数

  B. 虚函数不得是静态成员函数,但可以是友元函数

  C. 只能通过指针或引用调用虚函数

  D. 抽象类中的成员函数不一定都是虚函数

【答】 D

(21) 下列有关继承和派生的叙述中,正确的是(   )。

  A. 如果一个派生类私有继承其基类,则该派生类对象不能访问基类的保护成员

  B. 派生类的成员函数可以访问基类的所有成员

  C. 基类对象可以赋值给派生类对象

  D. 如果派生类没有实现基类的一个纯虚函数,则该派生类是一个抽象类

【答】 D

(22) 有以下程序:

```
#include <iostream>
```

```
using namespace std;
class Base
{
public:
    void fun1(){cout<<"Base\n";}
    virtual void fun2(){cout<<"Base\n";}
};
class Derived:public Base
{
public:
    void fun1(){cout<<"Derived\n";}
    void fun2(){cout<<"Derived\n";}
};
void f(Base& b){b.fun1();b.fun2();}
int main()
{
    Derived obj;
    f(obj);
    return 0;
}
```

执行这个程序,输出结果是(　　)。

  A. Base   B. Base   C. Derived   D. Derived

    Base   Derived   Derived   Base

【答】 B

(23) 有以下程序：

```
#include<iostream>
using namespace std;

class A{
public:
    virtual  void func1(){cout<<"A1";}
    void func2(){cout<<"A2";}
};
class B:public A{
public:
    void func1(){cout<<"B1";}
    void func2(){cout<<"B2";}
};
int main()
{
    A  *p = new B;
    p->func1()
    p->func2();
    return 0;
}
```

运行程序,屏幕上将显示输出(　　)。

  A. B1B2   B. A1A2   C. B1A2   D. A1B2

【答】 C

(24) 如果不使用多态机制,那么通过基类的指针虽然可以指向派生类对象,但只能访问从基类继承的成员,如有以下程序,没有使用多态机制。

```cpp
#include<iostream>
using namespace std;
class Base{
int a,b;
public:
    Base(int x , int y) { a = x; b = y;}
    void show(){ cout << a <<','<< b << endl;}
};

class Derived:public Base {
    int c,d;
public:
    Derived(int x,int y,int z,int m):Base(x,y)
    {c = z; d = m;}
    void show(){ cout << c <<','<< endl;}
};

int main()
{
    Base B1(50,50), * pb;
    Derived D1(10,20,30,40);
    pb = &D1;
    pb -> show();
    return 0;
}
```

运行时输出的结果是(　　)。
　　A. 10,20　　　　B. 30,40　　　　C. 20,30　　　　D. 50,50

【答】　A

3. 简答题

(1) 在 C++ 中能否声明虚构造函数？为什么？能否声明虚析构函数？为什么？

【答】　在 C++ 中不能声明虚构造函数,因为当开始调用构造函数时对象还未完成实例化,只有在构造完成后对象才能成为一个类的名副其实的对象,但析构函数可以是虚函数,而且通常声明为虚函数,即虚析构函数。

(2) 什么是抽象类？抽象类有何作用？可以声明抽象类的对象吗？为什么？

【答】　抽象类是一种特殊的类,是为了抽象目的而建立的,建立抽象类,就是为了通过它多态地使用其中的成员函数,为一个类族提供统一的操作界面。抽象类处于类层次的上层,一个抽象类自身无法实例化,也就是说我们无法声明一个抽象类的对象,而只能通过继承机制生成抽象类的非抽象派生类,然后再实例化。

(3) 什么是虚函数？空虚函数有何作用？定义虚函数有何限制？

【答】　虚函数是为实现某种功能而假设的函数,它是使用 virtual 关键字修饰的成员函数。空虚函数本身不执行任何动作,但可以保证建立一条从基类到派生类的虚函数路径,使派生类可以通过动态联编正确访问虚函数。

只有类的成员函数才可以是虚函数,静态成员不能是虚函数,构造函数不能是虚函数。

(4) 多态性和虚函数有何作用?

【答】 虚函数(virtual function)允许函数调用和函数体之间的联系在运行时才建立,是实现动态联编的基础。动态联编则在程序运行的过程中根据指针与引用实际指向的目标调用对应的函数,也就是在程序运行时才决定如何动作。虚函数经过派生之后可以在类族中实现运行时的多态,充分体现了面向对象程序设计的动态多态性。

(5) 什么是纯虚函数? 它有何作用?

【答】 纯虚函数(pure virtual function)是一个在基类中说明的虚函数,它在该基类中没有定义具体实现,要求各派生类根据实际需要定义函数实现。纯虚函数的作用是为派生类提供一个一致的接口。

抽象类描述的是所有派生类的高度抽象的共性,这些高度抽象、无法具体化的共性由纯虚函数来描述。带有纯虚函数的类被称为抽象类,一个抽象类至少具有一个纯虚函数。

4. 程序填空题

(1) 下面是类 fraction(分数)的定义,其中重载的运算符<<以分数形式输出结果,例如将三分之二输出为 2/3。在横线处填上适当的内容。

```
class fraction {
    int den;        //分子
    int num;        //分母
    friend ostream& operator <<(ostream&,fraction);
    ⋮
};
ostream& operator <<(ostream& os,fraction fr){
    _____①_____ ;
    return _____②_____ ;
}
```

【答】 ① os << fr.den <<"/"<< fr.num

② os

(2) 在下面程序的横线处填上适当的内容,使其输出结果为"0,56,56"。

```
#include <iostream>
using namespace std;
class base{
    public:
        _____①_____ func(){return 0;}
};
class derived:public base{
    public:
        int a,b,c;
        _____②_____ setValue(int x,int y,int z){a=x;b=y;c=z;}
        int func(){return(a+b)*c;}
};
int main()
{
    base b;
    derived d;
    cout << b.func()<<",";
    d.setValue(3,5,7);
```

```cpp
    cout << d.func()<<",";
    base& pb = d;
    cout << pb.func()<< endl;
    return 0;
}
```

【答】　① virtual int
　　　② void

(3) 在下面程序的横线处填上适当的内容，使该程序输出结果为：

Creating B
end of B
end of A

```cpp
#include <iostream>
using namespace std;
class A
{
    public:
        A(){}
        _____①_____ {cout <<"end of A"<< endl;}
};
class B:public A
{
    public:
        B(){_____②_____}
        ~B(){cout <<"end of B"<< endl;}
};
int main()
{
    A * pa = new B;
    delete pa
    return 0;
}
```

【答】　① virtual ~A()
　　　② cout <<"Creating B"<< endl;

(4) 下列程序的输出结果为 2，请将程序补充完整。

```cpp
#include <iostream>
using namespace std;
class Base
{
public:
    _____ void fun(){ cout << 1; }
};
class Derived:public Base
{
public:
    void fun() { cout << 2; }
};
int main()
```

```
{
    Base  *p = new Derived;
    p->fun();
    delete p;
    return 0;
}
```

【答】 virtual

（5）在下面程序中的横线处填上适当内容，使程序完整。

```
#include <iostream>
using namespace std;
class vehicle
{
    protected:
        int size;
        int speed;
    public:
        void setSpeed(int s){speed=s;}
        _____①_____ getSpeedLevel(){return speed/10;}
};
class car:public vehicle
{
    public:
        int getSpeedLevel(){return speed/5;}
};
class truck:public vehicle
{
    public:
        int getspeedLevel(){return speed/15;}
};
int maxSpeedLevel(vehicle_____②_____,vehicle_____③_____)
{
    if(v1.getSpeedLevel()>v2.getSpeedLevel())
        return 1;
    else
        return 2;
}
int main()
{
    turck t;
    car c;
    t.setSpeed(130);
    c.setSpeed(60);
    cout<<maxSpeedLevel(t,c)<<endl;         //此结果输出为 2
    return 0;
}
```

【答】 ① virtual int
② &v1
③ &v2

5. 程序分析题（写出运行结果）

(1)

```cpp
#include <iostream>
using namespace std;
class A {
    public:
    virtual void func(){cout <<"func in class A"<< endl;}
};
class B {
    public:
    virtual void func(){cout <<"func in class B"<< endl;}
};
class C: public A, public B {
    public:
    void func(){cout <<"func in class C"<< endl;}
};
int main()
{
    C c;
    A& pa = c;
    B& pb = c;
    C& pc = c;
    pa.func();
    pb.func();
    pc.func();
    return 0;
}
```

【答】 运行结果：

```
func in class C
func in class C
func in class C
```

(2)

```cpp
#include <iostream>
using namespace std;
class A{
public:
    virtual ~A(){
        cout <<"A::~A() called "<< endl;
    }
};
class B:public A{
    char *buf;
public:
    B(int i){ buf = new char[i]; }
    virtual ~B(){
        delete []buf;
        cout <<"B::~B() called"<< endl;
```

```
        }
};
void fun(A  * a) {
    delete a;
}
int main()
{   A   * a = new B(10);
    fun(a);
    return 0;
}
```

【答】 运行结果：

```
B::~B() called
A::~A() called
```

（3）

```
#include <iostream>
using namespace std;
class ONE{
    public:
        virtual void f(){cout<<"1";}
};
class TWO:public ONE{
    public:
        TWO(){cout<<"2";}
};
class THREE:public TWO{
    public:
        virtual void f() {TWO::f();cout<<"3";}
};
int main()
{
    ONE aa, * p;
    TWO bb;
    THREE cc;
    p = &cc;
    p->f();
    return 0;
}
```

【答】 运行结果：

```
2213
```

（4）

```
#include <iostream>
using namespace std;
class counter{
    private:
```

```cpp
        unsigned value;
    public:
        counter(){value = 0;};
        counter(unsigned int a){value = a;};
        counter&operator++();
        void display();
};
counter&counter::operator++(){
    value++;
    return  *this;
}
void counter::display(){
    cout <<"Total is "<< value << endl;
}
int main()
{   counter i(0),n(10);
    i = ++n;
    i.display();
    n.display();
    i = n++;
    i.display();
    n.display();
    return 0;
}
```

【答】 运行结果：

```
Total is 11
Total is 11
Total is 12
Total is 12
```

6. 编程题

(1) 重载运算符"<<"，使之能够使用 cout 将 Date 类对象的值以日期格式输出。

【答】 程序如下：

```cpp
#include < iostream >
using namespace std;
class Date                                    //定义日期类 Date
{
public:
    Date(int y = 1996, int m = 1, int d = 1);
    void NewDay();
    int days(int year, int month);
    void display()
    {
        cout << year <<" - "<< month <<" - "<< day << endl;
    }
    //<<运算符只能重载为类的非成员函数
    friend ostream &operator <<(ostream & output, const Date & d)
```

```cpp
        {
            output << d.year <<" - "<< d.month <<" - "<< d.day << endl;
            return output;
        }
private:
    int year;                                //年
    int month;                               //月
    int day;                                 //日
};

int Date::days(int year, int month)
    {
        bool leap;
        if((year % 400 == 0)||(year % 4 == 0&&year % 100!= 0))
            leap = true;
        else
            leap = false;
        switch(month)
        {
          case 1:
          case 3:
          case 5:
          case 7:
          case 8:
          case 10:
          case 12:
                return 31;
          case 4:
          case 6:
          case 9:
          case 11:
                return 30;
          case 2:
                if(leap)                                //是否闰月
                  return 29;
                else
                  return 28;
                break;
        }
    }
Date::Date(int y,int m,int d)
    {
    if(m > 12||m < 1){
        cout <<"Invalid month! "<< endl;
        m = 1;
    }
    if(d > days(y,m)) {
        cout <<"Invalid day! "<< endl;
        d = 1;
    }
    day = d;
    year = y;
```

```
            month = m;
        }
void Date::NewDay()
    {
        if (day < days(year,month))
            day++;
        else {
            day = 1;
            month++;
            if(month == 13) {
                year++;
                month = 1;
            }
        }
    }

int main()
{
    Date d1(2009,6,31);
    cout << d1;
    return 0;
}
```

运行结果：

```
Invalid day!
2009 - 6 - 1
```

(2) 定义一个 Location 类，重载运算符"＋"和"－"，实现平面位置的移动。

【答】 程序如下：

```
#include <iostream>
using namespace std;
class Location                                    //定义类 Location
{
public:
    Location(int xx, int yy)
    {
        x = xx;
        y = yy;
    }
    Location &operator + (Location &offset);      //重载运算符 +
    Location &operator - (Location &offset);      //重载运算符 -
    int getx(){ return x;}
    int gety(){ return y;}
private:
    int x;
    int y;
};
```

```
Location & Location::operator + (Location & offset)
{
    x = x + offset.x;
    y = y + offset.y;
    return * this;
}
Location & Location::operator - (Location & offset)
{
    x = x - offset.x;
    y = y - offset.y;
    return * this;
}

int main()
{
    Location p1(10,20),off(5,5);
    cout <<"("<< p1.getx()<<","<< p1.gety()<<")"<< endl;
    p1 = p1 + off;
    cout <<"("<< p1.getx()<<","<< p1.gety()<<")"<< endl;
    p1 = p1 - off;
    cout <<"("<< p1.getx()<<","<< p1.gety()<<")"<< endl;
    return 0;
}
```

运行结果：

```
(10,20)
(15,25)
(10,20)
```

(3) 重载操作符，实现集合类 Set 的操作，即包含于<=、并|、交 &、差-、增加元素+=、删除元素-=。

【答】 程序如下：

```
#include <iostream>
using namespace std;
class Set                               //集合类 Set
{
private:
    int n;                              //元素个数
    int *pS;                            //集合
public:
    Set() { n = 0; pS = NULL; }         //符合习惯,集合下标从1开始有效
    Set (Set & s)
    {
        n = s.n;
        if (n!= 0)
        {
```

```cpp
            pS = new int[n + 1];
            for (int i = 1; i <= n; i++)
                pS[i] = s.pS[i];
        }
    }
    ~Set()
    {
        if (pS)
        {
            delete []pS;
            pS = NULL;
            n = 0;
        }
    }

    void ShowElement()const                    //显示集合元素
    {
        cout <<"{ ";
        for (int i = 1; i < n; i++)
            cout << pS[i]<<", ";
        cout << pS[n]<<" }"<< endl;
    }
    bool IsEmpt()   { return n?false:true;}    //判断集合是否为空
    int size()  {       return n;}             //返回元素个数
    bool IsElement (int e)const;               //判断 e 是否属于集合
    bool operator <= (const Set &s)const;      //this <= s 判断当前集合是否包含于集合 s
    bool operator == (const Set &s)const;      //判断集合是否相等
    Set& operator += (int e);                  //增加元素 e
    Set& operator -= (int e);                  //删除元素 e
    Set  operator|(const Set &s)const;         //集合并
    Set  operator&(const Set &s)const;         //集合交
    Set  operator - (const Set &s)const;       //集合差
};

bool Set::IsElement(int e)const                //判断 e 是否属于集合
{
    for (int i = 1; i <= n; i++)
        if (pS[i] == e)
            return true;
    return false;
}

bool Set::operator <= (const Set & s)const
{
    if (n > s.n)
        return false;
    for (int i = 1; i <= n; i++)
        if (!s.IsElement (pS[i]))
            return false;
    return true;
}
```

```cpp
bool Set::operator == (const Set & s)const
{
    if (n!= s.n)
        return false;
    if ((*this)<= s&&s<= (*this))
        return true;
    return false;
}

Set & Set::operator += (int e)                  //增加元素 e
{
    if (IsElement(e))   return *this;
    Set tempe;
    for (int i = 1;i<= n;i++)
        tempe += pS[i];
    n++;
    if (pS!= NULL)
        delete[ ]pS;
    pS = new int[n + 1];
    for (i = 1;i<n;i++)
        pS[i] = tempe.pS[i];
    pS[n] = e;
    return *this;
}

Set & Set::operator -= (int e)                  //删除元素 e
{
    if (!(IsElement(e)))
        return *this;
    for (int i = 1;i<= n;i++)
        if (pS[i] == e)
            break;
    for (int j = i;j<n;j++)
        pS[i] = pS[i + 1];
    n--;
    return (*this);
}

Set Set::operator | (const Set & s)const        //集合并
{
    Set Rs;
    for (int i = 1;i<= n;i++)                   //先用当前的集合初始化结果集
        Rs += pS[i];
    for (i = 1;i<= s.n;i++)
        if (!IsElement(s.pS[i]))
            Rs += s.pS[i];
    return Rs;
}
```

```cpp
Set Set::operator & (const Set & s)const          //集合交
{
    Set Rs;
    for (int i = 1; i <= s.n; i++)
        if (IsElement(s.pS[i]))
            Rs += s.pS[i];
    return Rs;
}

Set Set::operator - (const Set & s)const          //集合差
{
    Set Rs;
    for (int i = 1; i <= n; i++)
        if (!(s.IsElement(pS[i])))
            Rs += pS[i];
    return Rs;
}

void main()
{
    Set s1;
    s1 += 1; s1 += 2; s1 += 3;                    //增加元素 +=
    cout <<"s1 = ";
    s1.ShowElement();
    Set s2;
    s2 += 1; s2 += 3; s2 += 5; s2 += 6;
    cout <<"s2 = ";
    s2.ShowElement();
    cout <<"s1 == s2?"<<(s1 == s2)<< endl;        //集合相等 ==
    s2 -= 5;                                      //删除元素 -=
    cout <<"s2 - 5 = ";
    s2.ShowElement();
    Set s3(s1);
    cout <<"s3 = ";
    s3.ShowElement();
    cout <<"s1 == s3?"<<(s1 == s3)<< endl;
    s3 += 9;
    cout <<"s3 + 9 = ";
    s3.ShowElement();
    cout <<"s1 <= s3?"<<(s1 <= s3)<< endl;        //集合包含 <=
    cout <<"s1|s2 = ";
    (s1|s2).ShowElement();                        //集合并 |
    cout <<"s1&s2 = ";
    (s1&s2).ShowElement();                        //集合交 &
    cout <<"s1 - s2 = ";
    (s1 - s2).ShowElement();                      //集合差 -
}
```

运行结果：

```
s1 = { 1, 2, 3 }
s2 = { 1, 3, 5, 6 }
s1 == s2?0
s2 - 5 = { 1, 3, 6 }
s3 = { 1, 2, 3 }
s1 == s3?1
s3 + 9 = { 1, 2, 3, 9 }
s1 <= s3?1
s1|s2 = { 1, 2, 3, 6 }
s1&s2 = { 1, 3 }
s1 - s2 = { 2 }
```

（4）定义一个长数据类 LongNum，能实现 LongNum 型数之间、LongNum 型数与 int 型数的加法和减法运算。重载运算符<<实现 LongNum 型数的输出。

【答】 程序如下：

```cpp
#include <iostream>
using namespace std;
const int MAX = 200;                            //最大位数为 MAX
class LongNum
{
public:
    char oper[MAX];                             //存储数字串
    int len;                                    //位数
    LongNum (char *pstr = "") { strcpy (oper,pstr); len = strlen (oper);}
    LongNum operator + (LongNum );              //实现 LongNum 与 LongNum 之间的加法
    LongNum operator + (int);                   //实现 LongNum 与 int 之间的加法
    LongNum operator - (LongNum);               //实现 LongNum 与 LongNum 之间的减法
    LongNum operator - (int);                   //实现 LongNum 与 int 之间的减法
};

istream & operator >> (istream &input, LongNum &temp)
{
    cin >> temp.oper;
    temp.len = strlen(temp.oper);
    return input;
}

ostream & operator << (ostream & output, LongNum & temp)
{
    int i = 0;
    if (temp.oper[0] == '-')
    {
        cout <<" - "; i = 1;
    }
    while (i < temp.len&&temp.oper[i] == '0')            //去掉前导 0
```

```cpp
        i++;
        if (i < temp.len)
            cout << temp.oper + i;                      //输出后面的数字串
        else cout << 0;
        return output;
}

LongNum LongNum::operator + (LongNum oper2)             //两个大整数相加
{
    LongNum Res;
    Res.len = len > oper2.len?len:oper2.len;
    for (int i = Res.len + 1;i >= 0;i -- )              //先将第一个操作数复制到结果中
        Res.oper[i] = i > Res.len - len?oper[i - Res.len - 1 + len]:'0';
    char c = 0;                                         //记录进位
    for (i = Res.len;i >= 0;i -- )
    {
        char r;
        r = (i > Res.len - oper2.len) ?
(Res.oper[i] - '0' + oper2.oper[i - Res.len + oper2.len - 1] - '0' + c) :(Res.oper [i] - '0' + c);
        c = r > 9?1:0;                                  //为数字结果
        Res.oper[i] = (c > 0)?(r - 10 + '0'):(r + '0'); //换成字符
    }
    Res.len = strlen (Res.oper);
    return Res;
}

LongNum  LongNum::operator + (int ioper2)
{
    char stroper2[MAX];
    itoa (ioper2,stroper2,10);                          //将整数转换为字符串
    LongNum oper2(stroper2);
    return (( * this) + oper2);
}

LongNum LongNum::operator - (LongNum oper2)
{
    LongNum Res;
    char roper2[MAX];                                   //存储第二个操作数
    int rlen;
    int low = 0;
    if ((len > oper2.len)||(len == oper2.len&&strcmp(oper,oper2.oper)>= 0) )
        {                                               //第一个大,结果为正数
            strcpy(Res.oper,oper);
            Res.len = len;

            strcpy (roper2,oper2.oper);
            rlen = oper2.len;
        }
        else                                            //第二个大,结果为负数
        {
```

```cpp
            low = 1;
            Res.oper[0] = '-';
            strcpy(Res.oper + 1, oper2.oper);
            Res.len = oper2.len + 1;

            strcpy(roper2, oper);                    //第二个操作数为this
            rlen = len;
        }
        char b = 0;                                  //借位
        int j = rlen - 1;
        for (int i = Res.len - 1; i >= low; i--, j--)
        {
            char r;
            r = (j >= 0)?( Res.oper[i] - '0' - (roper2[j] - '0') - b ) : (Res.oper[i] -
                '0' - b);
            b = r < 0?1:0;
            Res.oper[i] = b > 0?(r + 10 + '0'):(r + '0');

        }
        return Res;
}

LongNum LongNum::operator - (int ioper2)
{
    char stroper2[MAX];
    itoa(ioper2, stroper2, 10);                      //将整数转换为字符串
    LongNum oper2(stroper2);
    return ((*this) - oper2);
}

void main()
{
    LongNum oper1;                                   //设定为正数的操作
    LongNum oper2;
    cout <<"Input two long number:"<< endl;
    cin >> oper1;
    cin >> oper2;

    cout << oper1 <<"  +  "<< oper2 <<"  =  ";
    cout <<(oper1 + oper2)<< endl;

    cout << oper1 <<"  +  10000  =  ";
    cout <<(oper1 + 10000)<< endl;

    cout << oper1 <<"  -  "<< oper2 <<"  =  ";
    cout <<(oper1 - oper2)<< endl;

    cout << oper1 <<"  -  10000  =  ";
    cout <<(oper1 - 10000)<< endl;
}
```

运行结果：

```
Input two long number:
55555555556666666666
98765432100123456789
55555555556666666666 + 98765432100123456789 =
154320987656790123455
55555555556666666666 + 10000 = 55555555556666676666
55555555556666666666 - 98765432100123456789 =
-43209876543456790123
55555555556666666666 - 10000 = 55555555556666656666
```

（5）有一个交通工具类 vehicle，将它作为基类派生小车类 car、卡车类 truck 和轮船类 boat，定义这些类并定义一个虚函数用来显示各类信息。

【答】 程序如下：

```cpp
#include<iostream>
using namespace std;
class Vehicle                                    //定义基类 Vehicle
{
public:
    Vehicle() {
        cout<<"Vehicle constructor..."<<endl;
    }
    ~Vehicle(){
        cout<<"Vehicle destructor..."<<endl;
    }
    virtual void display() const = 0;
};

class Car :public Vehicle                        //定义派生类 Car
{
public:
    Car() {
        cout<<"Car constructor..."<<endl;
    }
    ~Car(){
        cout<<"Car destructor..."<<endl;
    };
    void display() const
    {
        cout<<"This is a car!"<<endl;
    }
};

class Truck :public Vehicle                      //定义派生类 Truck
{
public:
    Truck(){
        cout<<"Truck constructor..."<<endl;
    }
    ~Truck(){
```

```cpp
        cout <<"Truck destructor..."<< endl;
    }
    void display() const{
        cout <<"This is a truck!"<< endl;
    }
};

class Boat :public Vehicle                    //定义派生类 Boat
{
public:
    Boat(){
        cout <<"Boat constructor..."<< endl;
    }
    ~Boat(){
        cout <<"Boat destructor..."<< endl;
    }
    void display() const{
        cout <<"This is a Boat!"<< endl;
    }
};

int main()
{
    Vehicle *V;
    V = new Car;
    V -> display();
    delete V;

    V = new Truck;
    V -> display();
    delete V;

    V = new Boat;
    V -> display();
    delete V;
    return 0;
}
```

运行结果：

```
Vehicle constructor...
Car constructor...
This is a car!
Vehicle destructor...
Vehicle constructor...
Truck constructor...
This is a truck!
Vehicle destructor...
Vehicle constructor...
Boat constructor...
This is a Boat!
Vehicle destructor...
```

(6) 定义一个 shape 抽象类，派生出 Rectangle 类和 Circle 类，计算各派生类对象的面积 Area()。

【答】 程序如下：

```cpp
#include <iostream>
using namespace std;
class Shape                                    //定义抽象类 Shape
{
public:
    Shape(){}
    ~Shape(){}
    virtual double GetArea() = 0;              //纯虚函数
    virtual double GetPerimeter() = 0;         //纯虚函数
};

class Rectangle : public Shape                 //定义派生类 Rectangle
{
public:
    Rectangle(double L, double W){
        length = L;
        width = W;
    }
    ~Rectangle(){};
    double GetArea(){
        return length * width;
    }
    double GetPerimeter(){
        return 2 * length + 2 * width;
    }
private:
    double length;
    double width;
};

class Circle : public Shape                    //定义派生类 Circle
{
public:
    Circle(double r){
        radius = r;
    }
    ~Circle(){}
    double GetArea(){
        return 3.14 * radius * radius;
    }
    double GetPerimeter(){
        return 2 * 3.14 * radius;
    }
private:
    double radius;
};
```

```cpp
int main()
{
    Shape * s;
    s = new Rectangle(10,20);
    cout <<"The area of the Rectangle is "<< s->GetArea()<< endl;
    cout <<"The perimeter of the Rectangle is "<< s->GetPerimeter()<< endl;
    delete s;
    s = new Circle(10);
    cout <<"The area of the Circle is "<< s->GetArea()<< endl;
    cout <<"The perimeter of the Circle is "<< s->GetPerimeter()<< endl;
    delete s;
    return 0;
}
```

运行结果：

```
The area of the Rectangle is 200
The perimeter of the Rectangle is 60
The area of the Circle is 314
The perimeter of the Circle is 62.8
```

(7) 定义猫科动物类 Felid，由其派生出猫类（Cat）和豹类（Leopard），二者都包含虚函数 sound()，要求根据派生类对象的不同调用各自重载后的成员函数。

【答】 程序如下：

```cpp
#include <iostream>
using namespace std;
class Felid                                    //定义基类 Felid
{
public:
    Felid()
    {
        cout <<"Felid constructor..."<< endl;
    }
    ~Felid()
    {
        cout <<"Felid destructor..."<< endl;
    }
    virtual void sound() const
    {
        cout <<"Felid sound!"<< endl;
    }
};

class Cat :public Felid                        //定义派生类 Cat
{
public:
    Cat()
    {
```

```cpp
        cout <<"Cat constructor..."<< endl;
    }
    ~Cat()
    {
        cout <<"Cat destructor..."<< endl;
    };
    void sound() const
    {
        cout <<"Miaow !"<< endl;
    }
};

class Leopard :public Felid                    //定义派生类 Leopard
{
public:
    Leopard()
    {
        cout <<"Leopard constructor..."<< endl;
    }
    ~Leopard()
    {
        cout <<"Leopard destructor..."<< endl;
    }
    void sound() const
    {
        cout <<"Howl !"<< endl;
    }
};

int main()
{
    Felid * animal;
    animal = new Cat;
    animal -> sound();
    delete animal;
    animal = new Leopard;
    animal -> sound();
    delete animal;
    return 0;
}
```

运行结果：

```
Felid constructor...
Cat constructor...
Miaow !
Felid destructor...
Felid constructor...
Leopard constructor...
Howl !
Felid destructor...
```

1.8　习题8解答

1. 填空题

(1) 将相似的类和函数的<u>共同特性</u>（<u>或实现</u>）提取出来，用一种统一的方式予以描述，形成类模板和函数模板，这就是C++的<u>模板编程</u>（template programming）。

(2) <u>类模板</u>是能根据不同参数建立不同类型成员的类。

(3) C++中的模板是<u>逻辑功能</u>相同而<u>类型</u>不同的函数和类的一种抽象，是参数化的函数和类。

(4) <u>函数模板</u>可以用来创建一个通用功能的函数。

(5) 函数模板定义后，就可以用它生成各种具体的<u>模板函数</u>。

(6) 类模板就是带有类型参数的类，是能根据不同参数建立不同类型成员的具体类，这称为<u>类的实例化</u>。

(7) 在使用类模板建立对象时才根据给定的模板参数值<u>实例化</u>成具体的类。

(8) 当既存在重载函数，又有函数模板时，函数调用优先绑定<u>重载函数</u>，只有<u>不能精确匹配重载函数</u>时才实例化类模板。

2. 选择题

(1) 类模板的使用实际上是将类模板实例化成一个(　　)。
　　A. 函数　　　　　　B. 对象　　　　　　C. 类　　　　　　D. 抽象类
【答】C

(2) 类模板的模板参数(　　)。
　　A. 只能作为数据成员的类型　　　　B. 只可作为成员函数的返回类型
　　C. 只可作为成员函数的参数类型　　D. 以上3种均可
【答】D

(3) 类模板的实例化(　　)。
　　A. 在编译时进行　　　　　　B. 属于动态联编
　　C. 在运行时进行　　　　　　D. 在连接时进行
【答】A

(4) 类模板的参数(　　)。
　　A. 可以有多个　　　　　　B. 不能有基本数据类型
　　C. 可以是0个　　　　　　D. 参数不能给初值
【答】A

(5) 类模板实例化时的实参值(　　)。
　　A. 一定要和类模板的参数个数相同　　B. 不能是0个
　　C. 可以是0个　　　　　　　　　　　D. 可以多于类模板的参数个数
【答】C
【注解】　当类模板的模板参数均给出了初值时，类模板实例化时可以没有实参。

(6) 以下类模板的定义正确的是(　　)。
　　A. template < class T, int i=0 >　　　B. template < class T, class int i >
　　C. template < class T, typename T >　D. template < class T1, T2 >

【答】 A

(7) 以下类模板：

```
template < class T1, class T2 = int, int num = 10 >
class Tclass {...};
```

正确的实例化方式为(   )。

  A. Tclass < char &, char > C1; B. Tclass < char *, char, int > C1;
  C. Tclass < > C1; D. Tclass < char, 100, int >

【答】 A

【注解】 正确的实例化还依赖于模板参数在类中如何使用。

(8) 以下类模板，正确的实例化方式为(   )。

```
template < class T >
class TAdd {
    T x, y;
public:
    TAdd(T a, T b):x(a),y(b) { }
    int add() { return x + y; }
};
```

  A. TAdd < char > K; B. TAdd < double, double > K(3.4, 4.8);
  C. TAdd < char * > K('3', '4'); D. TAdd < float > K('3', '4');

【答】 D

(9) 采用以上类模板，下列程序运行的结果为(   )。

```
int main()
{
    TAdd < double > A(3.8, 4.8);
    TAdd < char > B(3.8,4.8);
    cout << A.add()<<","<< B.add();
    return 0;
}
```

  A. 8,8 B. 7,7 C. 7,8 D. 8,7

【答】 D

(10) 以下程序运行的结果为(   )。

```
template < class T >
class Num {
    T x;
public:
    Num() { }
    Num(T x) {this -> x = x;}
    Num < T > & operator + (const Num < T > & x2) {
        static Num < T > temp;
        temp.x = x + x2.x;
        return temp;
    }
    void disp() {cout <<"x = "<< x;}
};
```

```
    int main()
    {
        Num < int > A(3.8), B(4.8);
        A = A + B;
        A.disp();
        return 0;
    }
```

  A. x=7   B. x=8   C. x=3   D. x= 4

【答】 A

(11) 关于在调用模板函数时模板实参的使用,下列表述正确的是( )。

  A. 对于虚拟类型参数所对应的模板实参,如果能从模板函数的实参中获得相同的信息,则都可以省略

  B. 对于虚拟类型参数所对应的模板实参,如果它们是参数表中最后的若干个参数,则都可以省略

  C. 对于虚拟型参数所对应的模板实参,若能够省略则必须省略

  D. 对于常规参数所对应的模板实参,在任何情况下都不能省略

【答】 D

(12) 以下是函数模板定义的头部,正确的是( )。

  A. template < T >

  B. template < class T1,T2 >

  C. template < class t1, typename t2, int s=0 >

  D. template < class T1; class T2 >

【答】 C

(13) 若同时定义了以下 A、B、C、D 函数,fun(8,3.1)调用的是函数( )。

  A. template < class T1, class T2 > fun (T1, T2)

  B. fun (double, int)

  C. fun (char, float)

  D. fun (double, double)

【答】 A

3. 程序填空题

下面函数模板求 x^n,其中 n 为整数。

```
# include < iostream >
using namespace std;
    ①
double power(T x, int n)
{
    if(x == 0) return 0;
    if(n == 0) return 1;
        ②
    for( int i = 0; i < abs(n); i++)
        ③
    if(n < 0)
        return 1.0/powerx;
```

```
        else
            ④
}
```

【答】 ① template < class T >
② T powerx=1;
③ powerx=powerx * x;
④ return powerx;

4. 编程题

(1) 设计一个类模板,其中包括数据成员 T a[n]以及对其进行排序的成员函数 sort(),模板参数 T 可实例化成字符串。

【答】 如果 T 仅仅实例化成基本数据类型与 char * 类型的字符串,可通过重载 sort()实现,程序如下：

```cpp
#include <iostream>
using namespace std;
template <class T, int n>
class Array
{
  private:
      T a[n];
  public:
      void sort(double)                    //从小到大排序
      {
          for (int i = 0;i < n-1;i++)
              for (int j = i+1;j < n;j++)  //从待排序序列中选择一个最小的数组元素
              if (a[i]>a[j])
              {
                  T t;
                  t = a[i];                //交换数组元素
                  a[i] = a[j];
                  a[j] = t;
              }
      }
      void sort(char *)                    //重载 sort()对 char * 类型数据排序
      {
          for (int i = 0;i < n-1;i++)
              for (int j = i+1;j < n;j++)  //从待排序序列中选择一个最小的数组元素
              if (a[j] == NULL||(a[i]!= NULL&&strcmp(a[i],a[j])>0))
              {                            //短路求值避免 strcmp(NULL)
                  T t;
                  t = a[i];                //交换指针
                  a[i] = a[j];
                  a[j] = t;
              }
      }
      void disp()
      {
          for (int i = 0;i < n-1;i++)
```

```cpp
            if(a[i]!= NULL)                //避免对 NULL 输出
                cout << a[i]<<",\t";
            else
                cout <<",\t";
            if(a[i]!= NULL)
                cout << a[i]<< endl;
            else
                cout << endl;
        }
        Array(T a[])
        {
            for(int i = 0;i < n;i++)
                this -> a[i] = a[i];
        }
    };
int main()
{
    float f[] = {1.2,2.3,7.7,4,3.4,2.3};
    Array< float, 6 > a1(f);
    a1.sort(f[0]);
    a1.disp();
    char * a[6] = {"a","abc","ABC","abcd"};
    Array< char * ,6 > a2(a);
    a2.sort(a[0]);
    a2.disp();
    char c[] = {'a','b','C','2'};
    Array< char, 6 > a3(c);
    a3.sort(c[0]);
    a3.disp();
    return 0;
}
```

运行结果：

1.2,	2.3,	2.3,	3.4,	4,	7.7
,	,	ABC,	a,	abc,	abcd
,	,	2,	C,	a,	b

如果 T 的类型可以实例化成自定义的字符串类——String 类型,需要重载赋值运算符 =、插入运算符<<、比较运算符>,分别用于进行字符串的赋值、输出、比较。程序如下：

```cpp
# include < iostream >
//using namespace std;
using std::cout;
using std::endl;
using std::ostream;
class String {
    private:
        char * Str;
        int len;
```

```cpp
public:
    String()
    {
        len = 0;
        Str = NULL;
    }
    String(const char *p)
    {
        len = strlen(p);
        Str = NULL;
        if(len!= 0)
        {
            Str = new char[len + 1];
            strcpy(Str,p);
        }
    }
    String(String & r)
    {
        len = r.len;
        if(len!= 0)
        {
            Str = new char[len + 1];
            strcpy(Str,r.Str);
        }
    }
    ~String()
    {
        if (len > 0)
          delete [] Str;
    }
    String & operator = (const String & r) {     //重载=用于 String 类变量的赋值
        if(len > 0)
            delete [] Str;
        Str = NULL;
        len = r.len;
        if(len > 0)
        {
            Str = new char[len + 1];
            strcpy(Str,r.Str);
        }
        return * this;
    }
    bool operator > (const String & r) {         //重载>用于 String 类值的比较
        if (Str == NULL)
            return false;
        if (Str!= NULL&&r.Str == NULL)
            return true;
        int n = (len > r.len?len:r.len);
        for(int i = 0;i < n;i++) {               //两个 String 值均为空或均不为空
            if(Str[i]< r.Str[i])
                return false;
            if(Str[i]> r.Str[i])
```

```cpp
                return true;
            }
            return false;
        }
        friend ostream & operator <<(ostream &output, const String & s)
                                        //重载<<方便输出 String 类值
        {                               //<<运算符只能重载为类的非成员函数
            if(s.Str!= NULL)
                output << s.Str;
            return output;
        }
};
template < class T, int n >
class Array
{
    private:
        T a[n];
    public:
        void sort()                     //从小到大排序
        {
            for (int i = 0;i < n - 1;i++)
                for (int j = i + 1;j < n;j++)   //从待排序序列中选择一个最小的数组元素
                    if (a[i]> a[j])
                    {
                        T t;
                        t = a[i];                //交换数组元素
                        a[i] = a[j];
                        a[j] = t;
                    }
        }
        void disp()
        {
            for (int i = 0;i < n - 1;i++)
            cout << a[i]<<",\t";
            cout << a[i]<< endl;
        }
        Array(T a[])
        {
            for(int i = 0;i < n;i++)
                this -> a[i] = a[i];
        }
};
int main()
{
    float f[] = {1.2,2.3,7.7,4,3.4,2.3};
    Array < float, 6 > a1(f);
    a1.sort();
    a1.disp();
    String a[6] = {"a","abc","ABC","abcd"};
    Array < String,6 > a2(a);
    a2.sort();
    a2.disp();
    return 0;
}
```

运行结果:

```
1.2,    2.3,    2.3,    3.4,    4,     7.7
 ,       ,      ABC,    a,     abc,   abcd
```

(2) 设计一个类模板,其中包括数据成员 Ta[n]以及在其中查找数据元素的函数 int search(T),模板参数 T 可实例化成字符串。

【答】 本题的关键在于使用重载比较运算符==进行字符串比较。程序如下:

```cpp
#include <iostream>
//using namespace std;
using std::cout;
using std::endl;
using std::ostream;
class String {
    private:
        char *Str;
        int len;
    public:
    String()
    {
        len = 0;
        Str = NULL;
    }
    String(const char *p)
    {
        len = strlen(p);
        Str = NULL;
        if(len!= 0)
        {
            Str = new char[len+1];
            strcpy(Str,p);
        }
    }
    String(String & r)
    {
        len = r.len;
        if(len!= 0)
        {
            Str = new char[len+1];
            strcpy(Str,r.Str);
        }
    }
    ~String()
    {
        if (len > 0)
            delete[] Str;
    }
    String & operator = (const String & r) {    //重载=用于 String 类变量的赋值
        if(len > 0)
```

```cpp
                delete[] Str;
            Str = NULL;
            len = r.len;
            if(len > 0)
            {
                Str = new char[len + 1];
                strcpy(Str,r.Str);
            }
            return * this;
        }
        bool operator == (const String & r) {        //重载 == 用于 String 类值的比较
            if(len!= r.len)                          //长度不等
                return false;
            if(len == 0)
                return true;                         //两者均为空串
            for(int i = 0;i < len;i++)               //两者均不为空串
                if(Str[i]!= r.Str[i])
                    return false;
            return true;
        }
        //重载<<方便输出 String 类值(<<运算符只能重载为类的非成员函数)
 friend ostream & operator <<(ostream &output, const String & s)
    {
                if(s.Str!= NULL)
                    output << s.Str;
            return output;
        }
};

template < class T, int n >
class Array
{
  private:
        T a[n];
  public:
        int search(T key)                            //返回查找到的元素序号,没找到返回 -1
        {
          for (int i = 0;i < n;i++)
                if(a[i] == key)
                    return i;
            return -1;
        }
        void disp()
  {
            for (int i = 0;i < n-1;i++)
              cout << a[i]<<",\t";
            cout << a[i]<< endl;
        }
        Array(T a[])
        {
            for(int i = 0;i < n;i++)
                this -> a[i] = a[i];
```

```cpp
        }
};
int main()
{
    float f[] = {1.2,2.3,7.7,4,3.4,2.3};
    Array<float,6> a1(f);
    cout << a1.search(2.3) << endl;
    String a[6] = {"a","abc","ABC","abcd"};
    Array<String,6> a2(a);
    cout << a2.search("") << endl;
    return 0;
}
```

运行结果：

```
1
4
```

(3) 完善本章的 Student 类模板，使之可以添加、删除、查询学生记录，对学生的成绩进行排序。类模板如下：

```cpp
template <class TNO, class TScore, int num>
class Student
{
private:
    int n;                          //实际学生人数
    TNO StudentID[num];
    TScore score[num];
};
```

【答】 程序如下：

```cpp
#include <iostream>
using namespace std;
class String {
public:
    char Str[20];
    String()
    {
        Str[0] = '\0';
    }
    String(const char *p)
    {
        strcpy(Str,p);
    }
    String(const String &r)
    {
        strcpy(Str,r.Str);
    }
    String & operator = (const String & r)    //重载=用于 String 类变量的赋值
    {
```

```cpp
        strcpy(Str,r.Str);
        return *this;
    }
    bool operator == (const String & r) {     //重载 == 用于 String 类值的比较
        if(strcmp(Str,r.Str))
            return false;
        else
            return true;
    }
    //重载<<输出 String 类值(<<运算符只能重载为类的非成员函数)
    friend ostream & operator <<(ostream &output, const String & s)
    {
        output << s.Str;
        return output;
    }
};

template < class TNO, class TScore, int num >
class Student
{
private:
    int n;                              //实际学生人数
    TNO StudentID[num];
    TScore score[num];
public:
    void append(TNO ID, TScore s);
    void Delete(TNO ID);
    int  search(TNO);
    void sort();
    void DispAll();
    Student() {
        n = 0;
    }
};
template < class TNO, class TScore, int num >
void Student < TNO, TScore, num >::append(TNO ID, TScore s)
{
    if(n < num) {
        StudentID[n] = ID;
        score[n] = s;
        n++;
    }
}
template < class TNO, class TScore, int num >
void Student < TNO, TScore, num >::Delete(TNO ID)
{
    for(int i = 0; i < n; i++)
        if(StudentID[i] == ID) {
            for(int j = i;j < n;j++){
                StudentID[j] = StudentID[j + 1];
                score[j] = score[j + 1];
            }
            n--;
        }
```

```cpp
}
template < class TNO, class TScore, int num >
int Student < TNO, TScore, num >::search(TNO no)
{
    for(int i = 0; i < n; i++)
        if(StudentID[i] == no)
            return i + 1;
    return 0;
}
template < class TNO, class TScore, int num >
void Student < TNO, TScore, num >::sort()
{
    for (int i = 0; i < n - 1; i++)                            //按成绩降序排列
        for (int j = i + 1; j < n; j++)
            if (score[i] < score[j])
            {
                TScore ts;
                TNO tn;
                ts = score[i],        tn = StudentID[i];       //交换数组元素
                score[i] = score[j], StudentID[i] = StudentID[j];
                score[j] = ts,        StudentID[j] = tn;
            }
}
template < class TNO, class TScore, int num >
void Student < TNO, TScore, num >::DispAll()
{
    for(int i = 0; i < n; i++)
        cout << StudentID[i] << "\t" << score[i] << endl;
}
int main()
{
    Student < char * , float ,100 > group1;
    group1.append("202009125",70.5);
    group1.append("202009126",85);
    group1.append("202009110",69);
    group1.append("202009107",92);
    cout <<"Before sort:"<< endl;
    group1.DispAll();
    group1.sort();
    cout <<"After sort:"<< endl;
    group1.DispAll();
    Student < String, float ,100 > group2;
    String S1("202009125");
    String S2("202009126");
    group2.append(S1,70.5);
    group2.append(S2,85);
    group2.append("202009110",69);
    group2.append("202009107",92);
    cout <<"Before sort:"<< endl;
    group2.DispAll();
    group2.sort();
    cout <<"After sort:"<< endl;
    group2.DispAll();
    cout <<"After delete "<< S1 <<"\t"<< S2 << endl;
```

```
        group2.Delete(S1);
        group2.Delete(S2);
        group2.DispAll();
        cout << group2.search(S2);
        return 0;
}
```

运行结果：

```
Before sort:
202009125       70.5
202009126       85
202009110       69
202009107       92
After sort:
202009107       92
202009126       85
202009125       70.5
202009110       69
Before sort:
202009125       70.5
202009126       85
202009110       69
202009107       92
After sort:
202009107       92
202009126       85
202009125       70.5
202009110       69
After delete 202009125    202009126
202009107       92
202009110       69
0
```

（4）设计一个单向链表类模板，结点数据域中的数据从小到大排列，并设计插入、删除结点的成员函数。

【答】 为了节省内存空间，将头指针与当前结点指针设计成 static，为同类型链表所共有；为了便于操作，同样将链表的第一个结点当成头结点，不存储数据。链表的构成如图 1-2 所示。

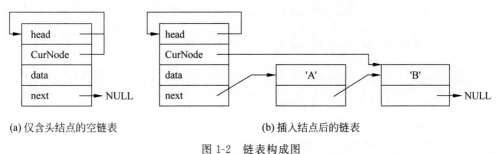

(a) 仅含头结点的空链表　　　　　(b) 插入结点后的链表

图 1-2　链表构成图

程序如下：

```cpp
#include <iostream>
using namespace std;
template <class TYPE>
class ListNode
{
private:
    TYPE data;
    ListNode * next;
    static ListNode * CurNode;
    static ListNode * head;
public:
    ListNode():next(NULL)
    {
        head = CurNode = this;
    }
    ListNode(TYPE NewData):data(NewData),next(NULL)
    { }
    void InsertNode(TYPE NewNode);
    void DeleteNode(TYPE NewNode);
    void DispList();
    void DelList();
};
template <class TYPE>
ListNode<TYPE> * ListNode<TYPE>::CurNode;
template <class TYPE>
ListNode<TYPE> * ListNode<TYPE>::head;

template <class TYPE>
void ListNode<TYPE>::InsertNode(TYPE NewData)
{
    ListNode * PreNode, * TempNode;
    PreNode = head;
    CurNode = PreNode->next;
    while(CurNode!=NULL&&CurNode->data<NewData)    //注意:短路求值
    {
        PreNode = CurNode;
        CurNode = CurNode->next;
    }
    if(CurNode == NULL)                             //添加
        PreNode->next = new ListNode(NewData);
    else {                                          //在 PreNode 后插入
        TempNode = new ListNode(NewData);
        TempNode->next = CurNode;
        PreNode->next = TempNode;
    }
}
template <class TYPE>
void ListNode<TYPE>::DeleteNode(TYPE NewData)
{
```

```cpp
    ListNode *PreNode;
    PreNode = head;
    CurNode = PreNode->next;
    while(CurNode!=NULL&&CurNode->data<NewData)    //注意:短路求值
    {
        PreNode = CurNode;
        CurNode = CurNode->next;
    }
    while(CurNode!=NULL&&CurNode->data==NewData)
    {//查找到,删除所有值为NewData的结点
        PreNode->next = CurNode->next;
        delete CurNode;
        CurNode = PreNode->next;
    }
}
template<class TYPE>
void ListNode<TYPE>::DispList()
{
    CurNode = head->next;
    while(CurNode!=NULL)
    {
        cout<<CurNode->data<<endl;
        CurNode = CurNode->next;
    }
}
template<class TYPE>
void ListNode<TYPE>::DelList()
{
    ListNode *q;
    CurNode = head->next;
    while(CurNode!=NULL)
    {
        q = CurNode->next;
        delete CurNode;
        CurNode = q;
    }
    head->next = NULL;
}
int main()
{
    ListNode<char> CList;
    CList.InsertNode('B');
    CList.InsertNode('A');
    CList.InsertNode('C');
    CList.InsertNode('A');
    CList.DispList();
    CList.DeleteNode('A');
    cout<<"after delete \'A\'"<<endl;
    CList.DispList();
    return 0;
}
```

运行结果：

```
A
A
B
C
after delete 'A'
B
C
```

(5) 利用 Stack 类模板计算不含括号的四则运算式的值。

【答】 程序如下：

```cpp
#include<iostream>
using namespace std;
template<class T>
class Stack
{
private:
    int size;
    int top;
    T* space;
public:
    Stack(int=10);
    ~Stack()    {
        delete [] space;
    }
    bool push(const T&);
    T pop();
    bool IsEmpty() const {
        return top==size;
    }
    bool IsFull() const{
        return top==0;
    }
};
template<class T>
Stack<T>::Stack(int size)
{
    this->size=size;
    space=new T[size];
    top=size;
}
template<class T>
bool Stack<T>::push(const T& element)
{
    if(!IsFull())
    {
        space[--top]=element;
```

```cpp
        return true;
    }
    return false;
}
template < class T >
T   Stack < T >::pop( )
{
    if(!IsEmpty())
        return space[top++];
    return 0;
}
template < class T >
T   str2num(char * str)                    //将字符串转换成数
{
    char tmpstr[20];
    strcpy(tmpstr,str);
    int i = 0;
    while(tmpstr[i]< = '9'&&tmpstr[i]> = '0'||tmpstr[i] == '.')
        i++;
    tmpstr[i] = '\0';
        return atof(tmpstr);
}
template < class T >
T ProcExp(char * p)
{
    T anum[20];                            //存储操作数
    char aop[20];                          //存储运算符
    int i = 0;                             //操作符个数
    Stack < T > sop(20);                   //存储运算符
    Stack < T > snum(20);                  //存储操作数
    T num1;
    //将操作数和运算符分离,分别存入数组 anum[ ]、aop[ ]中
    while( * p!= '\0') {
            anum[i] = str2num < T >(p);
                while( * p < = '9'&& * p > = '0'|| * p == '.')
                    p++;
                if( * p == '\0')
                    break;
                aop[i] = * p;
                p++;
                i++;
    }
    //进行运算式求值
    int j = 0;                             //指向操作符
    if(i > 0)
        snum.push(anum[j]);
    while (j < i) {
            switch (aop[j]){
            case ' * ':
                num1 = snum.pop();
                snum.push(num1 * anum[j + 1]);
                j++;
```

```
                break;
            case '/':
                num1 = snum.pop();
                snum.push(num1/anum[j + 1]);
                j++;
                break;
            case '+':
            case '-':
                if(!sop.IsEmpty())
                {
                    num1 = snum.pop();
                    if(sop.pop() == '+')
                        num1 = snum.pop() + num1;
                    else
                        num1 = snum.pop() - num1;
                    snum.push(num1);
                } else {
                  sop.push(aop[j]);
                  snum.push(anum[j + 1]);
                  j++;
                }
                break;
        }
    }
    if(!sop.IsEmpty())
            {
                num1 = snum.pop();
                if(sop.pop() == '+')
                    num1 = snum.pop() + num1;
                else
                    num1 = snum.pop() - num1;
                snum.push(num1);
    }
    return snum.pop();
}
int main()
{
    char frm[180];
    cin >> frm;
    cout << ProcExp< double >(frm)<< endl;
    cout << ProcExp< int >(frm)<< endl;
    return 0;
}
```

运行结果：

```
1 + 2/3 - 4/5 + 6 ↙
6.86667
7
```

【注解】 程序对表达式的合法性未作判断，因此，假定表达式是合法的。

（6）设计并测试一个描述队列（Queue）的类模板，该类模板模仿一个普通的等待队列：插入在线性结构的末尾进行，删除在该线性结构的另外一端进行。

【答】 程序如下：

```cpp
#include <iostream>
using namespace std;

//队列类模板定义
template <class T>
class Queue
{
   public:
       Queue(int s = 100)
       {
          size = s + 1;
          front = 0;
          rear = 0;
          data = new T[size];
       }
       ~Queue()
       {  delete []data; }
       void En_Queue(const T&x)            //进队列
       {
           data[rear++ % size] = x;
       }
       T DeQueue()                         //出队列
       {
           return data[front++ % size];
       };
       int IsEmpty()const                  //判断队列是否为空队列
       {
           return front == rear;
       }
       int IsFull() const                  //判断队列是否为满队列
       {
           return (rear + 1) % size == front;
       }
   private:
       int size;
       int front;
       int rear;
       T *data;
};

int main()
{
    Queue <char>  q(3);
    q.En_Queue('A');
    q.En_Queue('B');
    q.En_Queue('C');
    if(q.IsFull())
        cout <<"Queue is full."<< endl;
    else
```

```cpp
        cout <<"Queue is not full."<< endl;
    cout << q.DeQueue()<< endl;
    cout << q.DeQueue()<< endl;
    q.En_Queue('X');
    q.En_Queue('Y');
    while(!q.IsEmpty())
    {
        cout <<"Queue is not empty."<< endl;
        cout << q.DeQueue()<< endl;
    }
    cout <<"Queue is empty."<< endl;
    return 0;
}
```

运行结果：

```
Queue is full.
A
B
Queue is not empty.
C
Queue is not empty.
X
Queue is not empty.
Y
Queue is empty.
```

1.9 习题 9 解答

1. 填空题

（1）STL 中体现了泛型化程序设计的思想，它提倡使用现有的模板程序代码开发应用程序，是一种**代码重用**技术。

（2）**STL**（Standard Template Library）是 C++ 提供的标准模板库，它可以实现高效的泛型程序设计。

（3）STL 容器包括**顺序容器和关联容器**，利用容器适配器可以将顺序容器转换为新的容器。

（4）**顺序容器**（sequence container）以逻辑线性排列方式存储一个元素序列，容器类型中的对象在逻辑上被认为是在连续的存储空间中存储的。

（5）**关联容器**（associate container）中的数据元素不存储在顺序的线性数据结构中，它们提供了一个关键字（key）到值的关联映射。

（6）**容器适配器**就是将某个底层顺序容器转换为另一种容器，即以顺序容器作为数据存储结构，将其转换为一种某种特定操作特性的新容器。

（7）在 STL 中，**迭代器**如同一个特殊的指针，用于指向容器中某个位置的数据元素。

（8）在 STL 中，**函数对象**被广泛用作算法中子操作的参数，使算法变得更加通用。

2. 简答题

（1）面向对象程序设计与泛型程序设计有什么异同？

【答】 面向对象程序中更注重的是对问题域中数据的抽象，即所谓抽象数据类型（Abstract Data Type），而算法通常被附属于数据类型之中，独立于具体数据类型和数据结构；在面向对象程序设计中，算法往往被作为类的成员函数包含在类中。

在泛型程序设计中，算法与容器（数据结构）的分离具有通用性，不属于某个具体类。与针对问题和数据的面向对象的方法不同，泛型编程中强调的是通用的参数化算法，它们对各种数据类型和各种数据结构都能以相同的方式进行工作，从而实现源代码级的软件重用。例如，不管（容器）是数组、队列、链表还是堆栈，不管里面的元素（类型）是字符、整数、浮点数还是对象，都可以使用同样的（迭代器）方法遍历容器内的所有元素、获取指定元素的值、添加或删除元素，从而实现排序、检索、复制、合并等各种操作和算法。

（2）STL 编程是如何体现泛型程序设计思想的？

【答】 在 STL 编程中，大部分数据结构被抽象为通用的类模板，通过类型参数实例化为具体的类。算法则被抽象、被泛化，不仅独立于底层元素的类型，而且独立于所操作的容器，实现了算法与容器（数据结构）的分离，同一算法适用于不同的容器和数据类型，成为通用性算法，利用这些已经定义的算法和迭代器，程序设计人员可以方便、灵活地存取容器中存储的各种数据元素，从而实现泛型程序设计。

（3）什么是迭代器？其作用是什么？

【答】 在 STL 中，**迭代器**是一种广义的指针，用于指向容器中某个位置的数据元素，可以用来存取容器内存储的数据。迭代器是连接容器和算法的"纽带"，为数据提供了抽象，使写算法的人不必关心各种数据结构的细节。迭代器的功能是通过使用迭代器（一种通用的方法）来访问具有不同结构的各种容器中的每个元素。

3. 编程题

（1）利用向量容器装入整数 1~10，使用迭代器 iterator 和 accumulate 算法统计出这 10 个元素的和。

【答】 程序如下：

```cpp
#include <iostream>
#include <vector>
#include <numeric>
using namespace std;

int main()
{
    vector<int> v;
    int i;
    for(i=1;i<=10;i++){
        v.push_back(i);                    //在容器后端增加元素
    }
    //使用 iterator 顺序遍历所有元素
    for(vector<int>::iterator it=v.begin();it!=v.end();it++)
    {
        cout<<*it<<" ";                    //输出当前位置上的元素值
    }
```

```
        cout << endl;
        //统计并输出向量所有元素的和
        cout << accumulate(v.begin(),v.end(),0)<< endl;
        return 0;
}
```

运行结果：

```
1 2 3 4 5 6 7 8 9 10
55
```

(2) 利用 STL 算法和迭代器编程实现堆排序。

【答】 程序如下：

```
#include <iostream>
#include <algorithm>
#include <vector>
using namespace std;

int main() {
    const int SIZE = 10;
    int a[SIZE] = {10,3,8,11,20,7,19,5,16,1};
    vector <int> v(a,a+SIZE);
    ostream_iterator <int> output(cout," ");

    cout <<"堆排序之前: "<< endl;
    copy(v.begin(),v.end(),output);

    //创建堆
    make_heap(v.begin(),v.end());
    cout <<"\n建堆之后: "<< endl;
    copy(v.begin(),v.end(),output);

    //堆排序
    sort_heap(v.begin(),v.end());
    cout <<"\n堆排序之后: "<< endl;
    copy(v.begin(),v.end(),output);

    return 0;
}
```

运行结果：

```
堆排序之前：
10 3 8 11 20 7 19 5 16 1
建堆之后：
20 16 19 11 3 7 8 5 10 1
堆排序之后：
1 3 5 7 8 10 11 16 19 20
```

(3) 利用 STL 标准库函数实现有序值集合的包含、差、交、并等操作。

【答】 程序如下：

```cpp
#include <iostream>
#include <algorithm>
//#include <vector>
#include <set>
using namespace std;

int main() {
    const int SIZE1 = 10, SIZE2 = 5, SIZE3 = 20;
    int *ptr;
    int set1[SIZE1] = {0,1,2,3,4,5,6,7,8,9};
    int set2[SIZE2] = {1,2,3,4,5};
    int set3[SIZE2] = {1,2,3,11,12};
    ostream_iterator<int> output(cout," ");

    cout <<"set1 中的元素: "<< endl;
    copy(set1,set1 + SIZE1,output);

    cout <<"\nse21 中的元素: "<< endl;
    copy(set2,set2 + SIZE2,output);

    cout <<"\nset3 中的元素: "<< endl;
    copy(set3,set3 + SIZE2,output);

    //"includes(包含)"关系
    if(includes(set1,set1 + SIZE1,set2,set2 + SIZE2))
        cout <<"\nset1 包含 set2"<< endl;
    else
        cout <<"\nset1 不包含 set2"<< endl;

    if(includes(set1,set1 + SIZE1,set3,set3 + SIZE2))
        cout <<"\nset1 包含 set3"<< endl;
    else
        cout <<"\nset1 不包含 set3"<< endl;

    //"difference(差)"操作
    int difference[SIZE1];
    ptr = set_difference(set1,set1 + SIZE1,set2,set2 + SIZE2,difference);
    cout <<"\nset1 和 set2 的 difference: "<< endl;
    copy(difference,ptr,output);

    //"symmetric_difference(对称差)"操作
    int symmetric_difference1[SIZE1];
    ptr = set_symmetric_difference(set1,set1 + SIZE1,set2,set2 + SIZE2,symmetric_difference1);
    cout <<"\nset1 和 set2 的 symmetric_difference: "<< endl;
    copy(symmetric_difference1,ptr,output);

    int symmetric_difference2[SIZE1];
    ptr = set_symmetric_difference(set1,set1 + SIZE1,set3,set3 + SIZE2,symmetric_difference2);
    cout <<"\nset1 和 set3 的 symmetric_difference: "<< endl;
    copy(symmetric_difference2,ptr,output);
```

```
        //"intersection(交)"操作
        int intersection[SIZE1];
        ptr = set_intersection(set1,set1 + SIZE1,set2,set2 + SIZE2,intersection);
        cout <<"\nset1 和 set2 的 intersection: "<< endl;
        copy(intersection,ptr,output);


        //"union(并)"操作
        int set_union[SIZE3];
        ptr = std::set_union(set1,set1 + SIZE1,set3,set3 + SIZE2,set_union);
        cout <<"\nset1 和 set3 的 union: "<< endl;
        copy(set_union,ptr,output);

        cout << endl;
        return 0;
    }
```

运行结果：

```
    set1 中的元素：
    0 1 2 3 4 5 6 7 8 9
    se21 中的元素：
    1 2 3 4 5
    set3 中的元素：
    1 2 3 11 12
    set1 包含 set2

    set1 不包含 set3

    set1 和 set2 的 difference:
    0 6 7 8 9
    set1 和 set2 的 symmetric_difference:
    0 6 7 8 9
    set1 和 set3 的 symmetric_difference:
    0 4 5 6 7 8 9 11 12
    set1 和 set2 的 intersection:
    1 2 3 4 5
    set1 和 set3 的 union:
    0 1 2 3 4 5 6 7 8 9 11 12
```

（4）利用向量容器设计并实现一个图书评级类对若干图书进行评级，对评级后的图书可以进行删除、插入等操作。

【答】 程序如下：

```
#include <iostream>
#include <string>
#include <vector>
using namespace std;
```

```cpp
class Review {                              // 图书评价结构
private:
    string title;                           // 书名
    int rating;                             // 等级
public:
    bool FillReview(Review &rv);
    void ShowReview(const Review &rv);
};

bool Review::FillReview(Review &rv) {
    cout << "请输入书名(quit终止): ";
    cin >> rv.title;
    if (rv.title == "quit") return false;
    cout << "输入评级(0~5): ";
    cin >> rv.rating;
    if(!cin) return false;
    cin.get();
    return true;
}
void Review::ShowReview(const Review &rv) {
    cout << rv.rating << '\t' << rv.title << endl;
}

int main() {
    vector< Review > books;
    Review rv;

    while (rv.FillReview(rv)) books.push_back(rv);
    cout << "\n 输入图书评级如下: \n 评级\t 图书\n";

//人工循环
    int num = books.size();
    for (int i = 0; i < num; i++) rv.ShowReview(books[i]);
    cout << "\n 重复显示图书信息: \n 评级\t 图书\n";

//迭代器循环
    vector< Review >::iterator pr;
    for (pr = books.begin(); pr != books.end(); pr++)
        rv.ShowReview( *pr);
    vector< Review > oldrv(books);          //使用拷贝构造函数

//删除第 1 本图书
    if (num > 3) {
        books.erase(books.begin(), books.begin() + 1);
        cout << "\n 删除第 1 本图书后: \n";
        for (pr = books.begin(); pr != books.end(); pr++)
            rv.ShowReview( *pr);

//插入第 1 本图书到最后
        books.insert(books.end(), oldrv.begin(),oldrv.begin() + 1);
            cout << "\n 插入第 1 本图书到最后: \n";
        for (pr = books.begin(); pr != books.end(); pr++)
```

```
            rv.ShowReview( * pr);
      }

      return 0;
}
```

运行结果：

```
请输入书名(quit 终止)：C++编程思想
输入评级(0~5): 5
请输入书名(quit 终止)：泛型编程与 STL
输入评级(0~5): 4
请输入书名(quit 终止)：深度探索 C++对象编程
输入评级(0~5): 3
请输入书名(quit 终止)：C++标准程序库
输入评级(0~5): 2
请输入书名(quit 终止)：quit

输入图书评级如下：
评级      图书
5         C++编程思想
4         泛型编程与 STL
3         深度探索 C++对象编程
2         C++标准程序库

重复显示图书信息：
评级      图书
5         C++编程思想
4         泛型编程与 STL
3         深度探索 C++对象编程
2         C++标准程序库

删除第 1 本图书后：
4         泛型编程与 STL
3         深度探索 C++对象编程
2         C++标准程序库

插入第 1 本图书到最后：
4         泛型编程与 STL
3         深度探索 C++对象编程
2         C++标准程序库
5         C++编程思想
```

(5) 利用容器适配器建立容器对象分别为 vector 和 deque 的优先队列(priority_queue)，用于描述某公司所有员工的信息，并输出该公司中工资最高的员工的信息。

【答】 程序如下：

```
# include < iostream >
# include < queue >
```

```cpp
#include <string>
using namespace std;

class employee
{
public:
    employee() {}
    employee(long eID, string e_Name, float e_Salary);
    bool operator() (const employee& A, const employee& B)
    {
        return (A.salary < B.salary);
    }
    long ID;                            //员工编号
    string name;                        //员工姓名
    float salary;                       //员工薪水
};

employee::employee(long eID, string e_Name, float e_Salary):
        ID(eID), name(e_Name), salary(e_Salary) {}

//定义元素类型为employee、容器对象为vector的优先队列
typedef priority_queue<employee, vector<employee>, employee> V_QUEUE;
//定义元素类型为employee、容器对象为deque的优先队列
typedef priority_queue<employee, deque<employee>, employee> D_QUEUE;

int main()
{
    V_QUEUE v_employee;                 //容器对象为vector的优先队列对象
    D_QUEUE d_employee;                 //容器对象为deque的优先队列对象
    employee * temp;                    //employee对象指针

    //构造3个employee对象,插入容器对象为vector的优先队列
    v_employee.push(V_QUEUE::value_type(100312, "LiuWei", 4500));
    v_employee.push(V_QUEUE::value_type(1008, "HuCheng", 8000));
    v_employee.push(V_QUEUE::value_type(102009, "LiPeng", 6000));

    //动态空间申请,并用优先队列中优先级最高的元素初始化
    if(!v_employee.empty())
        temp = new employee(v_employee.top());
    //输出容器适配器对象为vector的优先队列的第一个元素,也就是优先级最高的元素(工资最高的
    //员工)
    cout << "The best-paid employee in the company:" << endl;
    cout << temp->ID << '\t' << temp->name << '\t' << temp->salary << endl;
    delete temp;
    //输出优先队列中元素的个数
    cout << "Is the priority_queue empty that the object of container is
        vector? " <<(v_employee.empty() ? "TRUE" : "FALSE") << endl;
    cout << "How many data elements are in the priority_queue that the object
        of container is vector? " << endl;
    cout << v_employee.size()   << endl;
```

```cpp
//构造 3 个 employee 对象,插入容器适配器对象为 deque 的优先队列
d_employee.push(D_QUEUE::value_type(100630, "TanFang", 5200));
d_employee.push(D_QUEUE::value_type(100220, "ZhangYong", 6800));
d_employee.push(D_QUEUE::value_type(1030, "Tony", 18000));
if(!d_employee.empty())
    temp = new employee(d_employee.top());
cout << "The best-paid employee in the company:" << endl;
cout << temp->ID << '\t' << temp->name << '\t' << temp->salary << endl;
delete temp;

//具有最高优先级的元素的出队列
cout <<"Delete The best priority element"<< endl;
d_employee.pop();
if(!d_employee.empty())
    temp = new employee(d_employee.top());
cout << "The best-paid employee in the firm : " << endl;
cout << temp->ID << '\t' << temp->name << '\t' << temp->salary << endl;
delete temp;

//输出优先队列中剩余元素的个数
cout << "Is the priority_queue empty that the object of container is deque?" << (d_employee.
    empty() ? " TRUE" : " FALSE") << endl;
cout << "How many data elements are in the priority_queue that the object of container is deque? "
    << endl;
cout << d_employee.size() << endl;
return 0;
}
```

运行结果:

```
The best-paid employee in the company:
1008    HuCheng 8000
Is the priority_queue empty that the object of container is Vector? FALSE
How many data elements are in the priority_queue that the object of container is vector?
3
The best-paid employee in the company:
1030    Tony    18000
Delete The best priority element
The best-paid employee in the firm :
100220  ZhangYong       6800
Is the priority_queue empty that the object of container is deque? FALSE
How many data elements are in the priority_queue that the object of container is deque?
2
```

(6) 用关联容器 multimap 设计并实现一个在线聊天程序中所有用户的"好友列表"类。每个用户可以有多个好友,其中键(key)是用户,值(value)是好友。容器中的每一项存储对应一个用户的好友,所有用户构成该容器的数据元素。

【答】 程序如下：

```cpp
#include <iostream>
#include <string>
#include <list>
#include <map>
using namespace std;

//创建好友列表
class FriendList
{
public:
    FriendList();
    void addFriend(const string &name, const string &partner);
    void removeFriend(const string &name, const string &partner);
    bool isFriend(const string& name, const string& partner) const;
    list<string> getfriends(const string &name) const;
protected:
    multimap<string,string> mfriends;
private:
    FriendList(const FriendList &src);
    FriendList &operator=(const FriendList &rhs);
};

FriendList::FriendList()
{
    cout <<"Create a FriendList."<< endl;
}

//判断是否好友
bool FriendList::isFriend(const string& name, const string& partner) const
{
    multimap<string, string>::const_iterator start,end;
    start = mfriends.lower_bound(name);
    end = mfriends.upper_bound(name);
    for(start; start != end; ++start)
    {
        if(start->second == partner)
        {
            return (true);
        }
    }
}

//加为好友
void FriendList::addFriend(const string &name, const string &partner)
{
    if(!isFriend(name, partner))
    {
        mfriends.insert(make_pair(name, partner));
    }
```

```cpp
}

//从好友名单中剔除
void FriendList::removeFriend(const string &name, const string &partner)
{
    multimap<string, string>::iterator start,end;
    start = mfriends.lower_bound(name);
    end = mfriends.upper_bound(name);
    for(start; start != end; ++start)
    {
        if(start->second == partner)
        {
            mfriends.erase(start);
            break;
        }
    }
}

//在好友名单中查找
list<string> FriendList::getfriends(const string &name) const
{
    pair<multimap<string, string>::const_iterator,multimap<string,
        string>::const_iterator> its;
    its = mfriends.equal_range(name);
    list<string> friends;
    for(its.first;its.first != its.second; ++its.first)
    {
        friends.push_back((its.first)->second);
    }
    return friends;
}

int main()
{
    FriendList goodfriends;
    goodfriends.addFriend("ZhouHao","ZhuXiaohua");
    goodfriends.addFriend("ZhouHao","HuChen");
    goodfriends.addFriend("ZhouHao","YuanXiaogang");
    goodfriends.addFriend("ZhouHao","LiNa");

    goodfriends.removeFriend("ZhouHao","YuanXiaogang");

    goodfriends.addFriend("HuChen","ZhouHao");
    goodfriends.addFriend("HuChen","LiFei");
    goodfriends.addFriend("HuChen","Tony");
    list<string> ZhouBuds = goodfriends.getfriends("ZhouHao");

    cout <<"ZhouHao's friends:"<< endl;
    for(list<string>::const_iterator it = ZhouBuds.begin();
        it!= ZhouBuds.end(); ++it)
    {
        cout <<"\t"<< *it << endl;
```

```
    }
    return 0;
}
```

运行结果:

```
Create a FriendList.
ZhouHao's friends:
        ZhuXiaohua
        HuChen
            LiNa
```

1.10 习题 10 解答

1. 填空题

(1) 标准输入流对象为 **cin**，与>>(**提取操作符**)连用，用于输入；**cout** 为标准输出流对象，与<<(**插入操作符**)连用，用于输出。

(2) 使用标准输入流对象 cin 和提取操作符>>连用进行输入时，将**空格**与**换行**当作分隔符，使用 **getline()** 成员函数进行输入时可以指定输入分隔符。

(3) 头文件 iostream 中定义了 4 个标准流对象，即 **cin**、**cout**、**cerr**、**clog**。

(4) 每一个输入/输出流对象都维护一个**格式状态字**，用它表示流对象当前的格式状态并控制流的格式。C++提供了使用**操纵符**与**格式状态字**来控制流的格式的方法。

(5) 格式控制的成员函数通过流对象调用；操纵符直接用在流中，但使用函数形式的操纵符要包含 **iomanip** 头文件。

(6) 在 ios 类中，除了提供控制数据流的格式标志、操纵符、成员函数外，还提供了流的错误侦测函数和错误状态位，用于标识流的状态，常用的错误侦测函数有 **good()**、**eof()**、**fail()**、**bad()**，对应的错误状态位为 **goodbit**、**eof bit**、**failbit**、**badbit**。

(7) 文件输入是指从文件向**内存**读入数据；文件输出则指从**内存**向文件输出数据。文件的输入/输出首先要**建立输入文件流，与打开的文件连接**；然后**从文件流中读入数据到内存**；最后**关闭文件流**。在打开文件、对文件读/写时要使用**是否成功的判断**保证文件操作的正确。

(8) 文本文件是存储 ASCII 码字符的文件，文本文件的输入可用>>(**提取操作符**)从输入文件流中提取字符实现。文本文件的输出可用<<(**插入操作符**)将字符插入到输出文件流来实现。

(9) 二进制文件是指含 ASCII 码字符外的数据的文件。二进制文件的输入/输出分别采用 read()、write() 成员函数，这两个成员函数的第一个参数的类型分别为 **char ***、**const char ***，如果实参类型不符，分别采用 **reinterpret_cast < char * >**、**reinterpret_cast < const char * >** 进行转换。

(10) 设定、返回文件读指针位置的函数分别为 **seekg()**、**tellg()**；设定、返回文件写指针位置的函数分别为 **seekp()**、**tellp()**。

2. 选择题

(1) 要进行文件的输出，除了包含头文件 iostream 外，还要包含头文件(　　)。

A. ifstream   B. fstream   C. ostream   D. cstdio

【答】 B

(2) 下列语句正确的是( )。

  A. cout << flags(ios::boolalpha)<< precision(10)<< 1.0/3.0 << endl;
  B. cout.setf(boolalpha);
  C. int i; cin >> width(10)>> fill('#')>> i;
  D. cout << setbase(16)<< 24 << endl;

【答】 D

(3) 执行以下程序：

```
char * str;
cin >> str;
cout << str;
```

若输入 abce　1234↙,则输出( )。

  A. abcd   B. abcd 1234   C. 1234   D. 输出乱码或出错

【答】 D

(4) 执行下列程序：

```
char a[200];
cin.getline(a,200,' ');
cout << a;
```

若输入 abcd　1234↙,则输出( )。

  A. abcd   B. abcd 1234   C. 1234   D. 输出乱码或出错

【答】 A

(5) 定义 char * p="abcd",能输出 p 的值("abcd"的地址)的是( )。

  A. cout << &p;   B. cout << p;
  C. cout <<(char *)p;   D. cout << const_cast < void * >(p);

【答】 D

(6) 定义"int a; int * pa=&a;",在下列输出式中,结果不是 pa 的值(a 的地址)的是( )。

  A. cout << pa;   B. cout <<(char *) pa;
  C. cout <<(void *)pa;   D. cout <<(int *) pa;

【答】 B

(7) 下列输出字符方式,错误的是( )。

  A. cout << put('A');   B. cout <<'A';
  C. cout.put('A');   D. char C='A';cout << C;

【答】 A

(8) 以下程序执行的结果是( )。

```
cout.fill('#');
cout.width(10);
cout << setiosflags(ios::left)<< 123.456;
```

  A. 123.456####   B. 123.4560000

C. ＃＃＃＃123.456　　　　　　　　D. 123.456

【答】 A

(9) 使用 ifstream 定义一个文件流,并将一个打开的文件与之连接,文件默认的打开方式为( )。

　　A. ios::in　　　B. ios::out　　　C. ios::trunc　　　D. ios::binary

【答】 A

(10) 使用 fstream 定义一个文件流,并将一个打开的文件与之连接,文件默认的打开方式为( )。

　　A. ios::in
　　B. ios::out
　　C. ios::in|ios::binary
　　D. ios::out|ios::binary

【答】 A

(11) 从一个文件中读一个字节存于 char c,正确的语句为( )。

　　A. file.read(reinterpret_cast<const char *>(&c),sizeof(c));
　　B. file.read(reinterpret_cast<char *>(&c),sizeof(c));
　　C. file.read((const char *)(&c),sizeof(c));
　　D. file.read((char *)c,sizeof(c));

【答】 B

(12) 将一个字符 char c='A'写到文件中,错误的语句为( )。

　　A. file.write(reinterpret_cast<const char *>(&c),sizeof(c));
　　B. file.write(reinterpret_cast<char *>(&c),sizeof(c));
　　C. file.write((char *)(&c),sizeof(c));
　　D. file.write((const char *)c,sizeof(c));

【答】 D

(13) 若文件的长度为 16 个字节,执行

```
myfile.seekg(-10, ios::end);
myfile.read((char *)(&c),sizeof(long)); myfile.tellg();
```

的返回值为( )。

　　A. 7　　　　　B. 10　　　　　C. 5　　　　　D. 2

【答】 B

(14) 读文件最后一个字节(字符)的语句为( )。

　　A. myfile.seekg(1,ios::end);
　　　c=myfile.get();
　　B. myfile.seekg(-1,ios::end);
　　　c=myfile.get();
　　C. myfile.seekp(ios::end,0);
　　　c=myfile.get();
　　D. myfile.seekp(ios::end,1);
　　　c=myfile.get();

【答】 B

(15) 语句 ofstream f("SALARY.DAT",ios_base::app)的功能是建立流对象 f,并试图打开文件 SALARY.DAT 与 f 关联,而且( )。

　　A. 若文件存在,将其置为空文件;若文件不存在,打开失败
　　B. 若文件存在,将文件指针定位于文件尾;若文件不存在,建立一个新文件

C. 若文件存在,将文件指针定位于文件首;若文件不存在,打开失败

D. 若文件存在,打开失败;若文件不存在,建立一个新文件

【答】 B

3. 程序填空题

(1)下列程序从键盘接受一行行文本,保存到文件中,输入空行表示输入结束,请填空。

```
#include <fstream>
#include <iostream>
using namespace std;
int main()
{
    char line[180];
    _____①_____ myfile;
    myfile.open("d:\\c++book\\lines.txt");
    if(_____②_____) {
        cerr <<"File open or create error!"<< endl;
        exit(1);
    }
    do {
        cin.getline(line,180);
        _____③_____
    }
    while(strlen(line)> 0&&!cin.eof());
    myfile.close();
    return 0;
}
```

【答】 程序如下:

```
#include <fstream>
#include <iostream>
using namespace std;
int main()
{
    char line[180];
    ofstream myfile;
    myfile.open("d:\\c++book\\lines.txt");
    if(!myfile) {
        cerr <<"File open or create error!"<< endl;
        exit(1);
    }
    do {
        cin.getline(line,180);
        myfile << line << endl;
    }
    while(strlen(line)> 0&&!cin.eof());
    myfile.close();
    return 0;
}
```

(2)下列程序读入 C++源程序,在每行的开头标明该行的长度,另存该文件。

```
#include <fstream>
```

```
#include <iostream>
using namespace std;
int main()
{
    char line[180];
    ifstream cppfile;
    ____①____ outfile;
    cppfile.open("d:\\c++book\\p11.cpp");
    outfile.open("d:\\c++book\\p11.txt");
    if(____②____) {
        cerr<<"File open error!"<< endl;
        exit(1);

    }
    if(____③____) {
        cerr<<"File create error!"<< endl;
        exit(1);
    }
    while(____④____) {
        ____⑤____ << strlen(line)<<" "<< line << endl;
    }
    cppfile.close();
    outfile.close();
    return 0;
}
```

【答】 程序如下：

```
#include <fstream>
#include <iostream>
using namespace std;
int main()
{
    char line[180];
    ifstream cppfile;
    ofstream outfile;
    cppfile.open("d:\\c++book\\p11.cpp");
    outfile.open("d:\\c++book\\p11.txt");
    if(!cppfile) {
        cerr<<"File open error!"<< endl;
        exit(1);

    }
    if(!outfile) {
        cerr<<"File create error!"<< endl;
        exit(1);
    }
    while(cppfile.getline(line,180)) {
        outfile << strlen(line)<<" "<< line << endl;
    }
    cppfile.close();
    outfile.close();
```

```
        return 0;
}
```

【注解】 ④处填 cppfile >> line 会将文本中的空格当成换行,因此是错误的。

(3) 下列程序将一个文件复制到另一个文件中,路径与文件名通过键盘输入。

```
#include <fstream>
#include <iostream>
using namespace std;
int main()
{
    char buff[1024];
    char source[80],target[80];
    long bytes;
    ifstream infile;
    ofstream outfile;
    cout <<"source file"<< endl;
    cin >> source;
    cout <<"target file"<< endl;
    cin >> target;
    infile.open(source,_____①_____);
    outfile.open(target,_____②_____);
    if(_____③_____) {
        cerr <<"File open error!"<< endl;
        exit(1);
    }
    if(_____④_____) {
        cerr <<"File create error!"<< endl;
        exit(1);
    }
    infile.seekg(0,ios::end);
    bytes = _____⑤_____;
    infile.seekg(0,ios::beg);
    infile.read(buff,bytes % 1024);
    outfile.write(buff,bytes % 1024);
    while(_____⑥_____) {
        _____⑦_____;
    }
    infile.close();
    outfile.close();
    return 0;
}
```

【答】 程序如下:

```
#include <fstream>
#include <iostream>
using namespace std;
int main()
{
    char buff[1024];
```

```cpp
        char source[80],target[80];
        long bytes;
        ifstream infile;
        ofstream outfile;
        cout <<"source file"<< endl;
        cin >> source;
        cout <<"target file"<< endl;
        cin >> target;
        infile.open(source,   ios::binary);
        outfile.open(target, ios::binary|ios::out|ios::trunc);
        if(!infile) {
            cerr <<"File open error!"<< endl;
            exit(1);
        }
        if(!outfile) {
            cerr <<"File create error!"<< endl;
            exit(1);
        }
        infile.seekg(0,ios::end);
        bytes = infile.tellg();
        infile.seekg(0,ios::beg);
        infile.read(buff,bytes % 1024);
        outfile.write(buff,bytes % 1024);
        while(infile.read(buff,1024)) {
            outfile.write(buff,1024);
        }
        infile.close();
        outfile.close();
        return 0;
    }
```

4. 编程题

（1）编写一个程序，将两个文件合并成一个文件。

【答】 程序如下：

```cpp
# include < fstream >
# include < iostream >
using namespace std;
int main()
{
    char buff[1024];
    char source1[80],source2[80],target[80];
    long bytes;
    ifstream infile1,infile2;
    ofstream outfile;
    cout <<"source file1"<< endl;
    cin >> source1;
    cout <<"source file2"<< endl;
    cin >> source2;
    cout <<"target file"<< endl;
```

```cpp
    cin >> target;
    infile1.open(source1, ios::binary);
    infile2.open(source2, ios::binary);
    outfile.open(target, ios::binary|ios::out|ios::trunc);
    if(!outfile) {
        cerr <<"File create error!"<< endl;
        exit(1);
    }
    if(infile1) {
        infile1.seekg(0,ios::end);
        bytes = infile1.tellg();
        infile1.seekg(0,ios::beg);
        infile1.read(buff,bytes % 1024);           //先读1k的零头
        outfile.write(buff,bytes % 1024);
        while(infile1.read(buff,1024))             //以1k为单位读写
            outfile.write(buff,1024);
        infile1.close();
    }
    if(infile2) {
        infile2.seekg(0,ios::end);
        bytes = infile2.tellg();
        infile2.seekg(0,ios::beg);
        infile2.read(buff,bytes % 1024);
        outfile.write(buff,bytes % 1024);
        while(infile2.read(buff,1024))
            outfile.write(buff,1024);
        infile2.close();
    }
    outfile.close();
    return 0;
}
```

(2) 编写一个程序,统计一篇英文文章中单词的个数与行数。

【答】 假定英文文章没有语法错误,单词为含26个大小写英文字母的字符串,每行用换行符隔开。程序如下:

```cpp
#include <fstream>
#include <iostream>
using namespace std;
int words(char * line)                            //统计一行文章中的单词数
                                                  //单词为含26个大小写英文字母的字符串
{
    int words = 0;
    bool PreChar = false;                         //前一个字符是否为英文字母
    bool CurChar = false;                         //当前字符是否为英文字母
    for(int i = 0;i < strlen(line);i++)
    {
        if(line[i]>= 'A'&&line[i]<= 'Z'||line[i]>= 'a'&&line[i]<= 'z')
            CurChar = true;
        else
            CurChar = false;
```

```cpp
        if(!PreChar&&CurChar)
            words++;
        PreChar = CurChar;
    }
    return words;
}
int main()
{
    const int N = 1024;
    char line[N];                                    //存储文章的一行
    int lines = 0, TotalWords = 0;
    char filename[40];
    ifstream txtfile;
    cout <<"text file:"<< endl;
    cin >> filename;
    txtfile.open(filename);
    if(!txtfile) {
        cerr <<"File open error!"<< endl;
        exit(1);
    }
    while(txtfile.getline(line,N)) {
        TotalWords += words(line);
        lines++;
    }
    txtfile.close();
    cout <<"total lines: "<< lines << endl;
    cout <<"total words: "<< TotalWords << endl;
    return 0;
}
```

(3) 编写一个程序,在 C++ 源程序的每行前加行号和一个空格。

【答】 程序如下:

```cpp
#include <fstream>
#include <iostream>
using namespace std;
int main()
{
    char line[180];
    char source[80],target[80];
    ifstream cppfile;
    ofstream outfile;
    cout <<"source file"<< endl;
    cin >> source;
    cout <<"target file"<< endl;
    cin >> target;
    cppfile.open(source);
    outfile.open(target);
    if(!cppfile) {
```

```
            cerr<<"File open error!"<<endl;
            exit(1);
        }
        if(!outfile) {
            cerr<<"File create error!"<<endl;
            exit(1);
        }
        int i = 1;
        while(cppfile.getline(line,180)) {
            outfile<<i<<" "<<line<<endl;
            i++;
        }
        cppfile.close();
        outfile.close();
        return 0;
    }
```

(4) 编写一个程序,输出 ASCII 码值从 20 到 127 的 ASCII 码字符表,格式为每行 10 个。

【答】 程序如下:

```
#include <fstream>
#include <iostream>
using namespace std;
int main()
{
    char line[180];
    char source[80],target[80];
    ifstream cppfile;
    ofstream outfile;
    cout<<"source file"<<endl;
    cin>>source;
    cout<<"target file"<<endl;
    cin>>target;
    cppfile.open(source);
    outfile.open(target);
    if(!cppfile) {
        cerr<<"File open error!"<<endl;
        exit(1);
    }
    if(!outfile) {
        cerr<<"File create error!"<<endl;
        exit(1);
    }
    int i = 1;
    while(cppfile.getline(line,180)) {
        outfile<<i<<" "<<line<<endl;
        i++;
    }
    cppfile.close();
    outfile.close();
```

```
        return 0;
}
```

文件内容为：

```
¶⊥ ┬ ┤ ↑ ├ → ←
 -!"# $ % &'
 ( ) * + , - . /01
23456789 : ;
< = >?@ABCDE
FGHIJKLMNO
PQRSTUVWXY
Z[\]^_`abc
defghijklm
nopqrstuvw
xyz{|}~
```

（5）重载<<和>>进行时间类型数据的输入/输出。

【答】 程序如下：

```cpp
#include <iostream>
using namespace std;
class Clock {
    private:
        int H,M,S;
    public:
        void SetTime(int h,int m,int s)
        {
            H = (h>=0&&h<24)?h:0;
            M = (m>=0&&m<60)?m:0;
            S = (s>=0&&s<60)?s:0;
        }
    void ShowTime()
    {
        cout << H <<":"<< M <<":"<< S << endl;
    }
    //重载>>输入 Clock 类值(>>运算符只能重载为类的非成员函数)
    friend istream & operator >>(istream &input, Clock & c)
    {
        input >> c.H;
        input.ignore();                    //忽略":"
        input >> c.M;
        input.ignore();                    //忽略":"
        input >> c.S;
        return input;
    }
    //重载>>输出 Clock 类值
    friend ostream & operator <<(ostream &output, const Clock & c)
```

```cpp
        {
            output << c.H <<":"<< c.M <<":"<< c.S << endl;
            return output;
        }
};

int main()
{
    Clock c1;
    cin >> c1;
    cout << c1;
    return 0;
}
```

运行结果：

```
21:30:45 ↙
21:30:45
```

(6) 定义一个 Student 类，其中包含学号、姓名、成绩数据成员。建立若干个 Student 类对象，将它们保存到文件 Record.dat 中，然后显示文件中的内容。

【答】 程序如下：

```cpp
#include <iostream>
#include <fstream>
using namespace std;
class Student
{
    private:
        long No;                                        //学号
        char *Name;                                     //姓名
        int  Score;                                     //成绩
    public:
        Student(long = 0, char * = NULL, int = 0);      //构造函数
        void ShowStudent();                             //显示学生信息
};
Student::Student(long no, char *name, int score)        //构造函数
{
        No = no;
        Name = name;
        Score = score;
}
void Student::ShowStudent()
{
        cout << No <<"\t"<< Name <<"\t"<< Score << endl;
}
int main()
{
```

```cpp
        Student S1[3] = {Student(202007001,"Li Yapeng",76),
                         Student(202007002,"Wang Fei",82),
                         Student(202007003,"Li Jian",95)};
    fstream objfile;
    objfile.open("d:\\c++book\\Students.dat",ios::in|ios::out|ios::
    binary|ios::trunc);
    if(!objfile) {
        cerr <<"File open error!"<< endl;
        exit(1);
    }
    for(int i = 0;i < 3;i++)
    {//将对象写入文件
        objfile.write((char * )&S1[i],sizeof(S1[i]));
    }
    Student S;
    objfile.seekg(0,ios::beg);                      //文件指针移至文件头
    objfile.read((char * )&S,sizeof(S));
    cout <<"Output object from file:"<< endl;
    while(!objfile.eof()) {                         //从文件中读入对象
        S.ShowStudent();
        objfile.read((char * )&S,sizeof(S));
    }
    return 0;
}
```

运行结果：

```
Output object from file:
202007001        Li Yapeng      76
202007002        Wang Fei       82
202007003        Li Jian        95
```

1.11　习题 11 解答

1. 选择题

(1) s0 是一个 string 类串，下列定义串 s1 错误的是(　　)。

　　A. string s1(3,"A")；　　　　　　B. string s1(s0,0,3)；

　　C. string s1("ABC",0,3)；　　　　D. string s1＝"ABC"；

【答】　A

【注解】　若 A 改为 string s1(3,'A')，则正确。

(2) char * S0＝"12345"，对 string 类串 s1 初始化错误的是(　　)。

　　A. string s1＝S0；　　　　　　　B. string s1(S0)；

　　C. string s1(S0, 0, 3)；　　　　D. string * s1＝S0；

【答】　D

(3) 求 string 类串 S 长度的表达式为(　　)。

A. S.capacity()　　B. sizeof(S)　　C. strlen(S)　　D. S.length()

【答】 D

(4) S 为 string 类对象，下列表达式编译错的是(　　)。

A. S.size()　　B. sizeof(S)　　C. strlen(S)　　D. S.length()

【答】 C

(5) S、T 为 string 类对象，下列表达式编译错的是(　　)。

A. S = T;　　B. S[1] = T[1];　　C. S - = T;　　D. S += T;

【答】 C

(6) 若定义：

```
string s = ("ABCDEF");
char  *q = "123456";
string::iterator p = s.begin();
```

下列表达式错误的是(　　)。

A. p=q;
B. *(p+1)= *(q+1);
C. cout << p.begin();
D. q[1]=p[1];

【答】 C

2. 编程题

(1) 输入一个句子，计算回文单词的个数。回文单词是指顺读、倒读都一样的单词，例如 noon 是一回文单词。

【答】 程序如下：

```cpp
#include <string>
#include <iostream>
using namespace std;
bool IsPalindrome(string str)
{
    string::iterator itr1 = str.begin();
    string::reverse_iterator itr2 = str.rbegin();
    int pos(0);
    while(pos < str.length()/2)
    {
        if(itr1[pos]!= itr2[pos])
            return false;
        pos++;
    }
    return true;
}
int main()
{
    string str,sstr;                    //str 存储句子、sstr 存储单词
    string::iterator itr1;
    int pos1(0),pos2(0);                //pos1 指向句子中的单词头、pos2 指向单词尾
    int num(0);
    cout <<"Input a string:";
    getline(cin,str);
```

```
        itr1 = str.begin();
        while(pos2 <= str.length())
        {
            if(itr1[pos2] == ' '||pos2 == str.length())
            {
                sstr = str.substr(pos1,pos2 - pos1);
                if(IsPalindrome(sstr))
                    num++;
                pos2++;
                pos1 = pos2;
            } else
                pos2++;
        }
        cout <<"Palindromes:"<< num << endl;
        return 0;
    }
```

运行结果：

```
Input a string:a bib dad bed noon adobe↵
↵
Palindromes:4
```

（2）编写一个程序，统计一篇英语文章中某个单词出现的次数。

【答】 程序如下：

```
#include <string>
#include <iostream>
using namespace std;
bool IsAlphabet(char c)                //是否字母
{
    if(c >= 'a'&&c <= 'z'||c >= 'A'&&c <= 'Z')
        return true;
    else
        return false;
}
int main()
{
    string text = "It was Sunday. I never get up early on Sundays. I sometimes
        stay in bed until lunch time. Last Sunday I got up very late. ";
    string word;                       //str 存储句子、word 存储单词
    string::iterator itr = text.begin();
    int pos1(0),pos2(0);               //pos1 指向文章头、pos2 指向找到的单词
    int num(0);
    cout <<"Input a word:";
    getline(cin,word);
    while((pos2 = text.find(word,pos1))!= string::npos)    //如果找到
    {
        if((pos2 == 0||!IsAlphabet(itr[pos2 - 1]))
```

```
            &&(pos2 + word.length() == text.length()||!IsAlphabet(itr[pos2 + word.
               length()])))
                  //如果找到的是一个单词
            { num++;
                pos1 = pos2 + word.length();
            }else                       //如果找到的是单词的一部分
                pos1++;
        }
        cout <<"Fine word \""<< word <<"\" "<< num <<" times"<< endl;
        return 0;
    }
```

运行结果:

```
Input a word:Sunday ↙
↙
Fine word "Sunday" 2 times
Input a word:I ↙
F ↙
ine word "I" 3 times
```

(3) 在一个串中查找一个子串,显示出现的次数,并将它们全部替换成新串。

【答】 程序如下:

```
#include <string>
#include <iostream>
using namespace std;
int main()
{
    string text = "It was Sunday. I never get up early on Sundays. I sometimes
       stay in bed until lunch time. Last Sunday I got up very late.";
    string OldStr,NewStr;
    string::iterator itr = text.begin();
    int pos1(0),pos2(0);              //pos1 指向文章头、pos2 指向找到的字符串
    int num(0);
    cout <<"Input a string:";
    cin >> OldStr;
    cout <<"Input a new string:";
    cin >> NewStr;
    while((pos2 = text.find(OldStr,pos1))!= string::npos)       //如果找到
    {
        text.replace(pos2, OldStr.length(), NewStr);
        num++;
        pos1 = pos2 + NewStr.length();
    }
    cout << num <<" string replaced"<< endl;
    cout <<"New text is:"<< text << endl;
    return 0;
}
```

运行结果：

```
Input a string:Sunday↙
Input a new string:Saturday↙
3 string replaced
New text is:It was Saturday. I never get up early on Saturdays.
 I sometimes stay in bed until lunch time. Last Saturday I got
up very late.
```

1.12　习题12解答

1. 选择题

(1) 下列关于异常的叙述错误的是(　　)。

　　A. 编译错属于异常，可以抛出

　　B. 运行错属于异常

　　C. 硬件故障也可当异常抛出

　　D. 只要是编程者认为是异常的都可当异常抛出

【答】　A

(2) 下列叙述错误的是(　　)。

　　A. throw 语句必须书写在 try 语句块中

　　B. throw 语句必须在 try 语句块中直接运行或通过调用函数运行

　　C. 一个程序中可以有 try 语句而没有 throw 语句

　　D. throw 语句抛出的异常可以不被捕获

【答】　A

(3) 关于函数声明 float fun(int a,int b) throw(),下列叙述正确的是(　　)。

　　A. 表明函数抛出 float 类型异常

　　B. 表明函数抛出任何类型异常

　　C. 表明函数不抛出任何类型异常

　　D. 表明函数实际抛出的异常

【答】　C

(4) 下列叙述错误的是(　　)。

　　A. catch(...)语句可捕获所有类型的异常

　　B. 一个 try 语句可以有多个 catch 语句

　　C. catch(...)语句可以放在 catch 语句组的中间

　　D. 程序中 try 语句和 catch 语句是一个整体，缺一不可

【答】　C

(5) 若定义"int a[2][3];",则使用表达式"a[2][3]=3;",下列叙述正确的是(　　)。

　　A. 数组下标越界，会引发异常

　　B. 数组下标越界，不引发异常

　　C. 数组下标没越界，不会引发异常

D. 数组下标越界,语法错误

【答】 B

(6) 下列程序运行的结果为(   )。

```cpp
#include <iostream>
using namespace std;
class S
{
public:
    ~S()
    {
        cout <<"S"<<"\t";
    }
};
char fun0() {
    S s1;
    throw('T');
    return '0';
}
int main()
{
    try
    {
        cout << fun0()<<"\t";
    }
    catch(char c)
    {
        cout << c <<"\t";
    }
    return 0;
}
```

A. S T   B. 0 S T   C. 0 T   D. T

【答】 A

2. 编程题

(1) 编写一个程序,演示异常嵌套处理时各层函数中局部对象构造与析构的过程。

【答】 程序如下:

```cpp
#include <iostream>
using namespace std;
class aclass
{
    int num;
public:
    aclass(int n) {
        num = n;
        cout <<"constructing..."<< num << endl;
    }
    ~aclass() {
```

```cpp
        cout <<"deconstructing..."<< num << endl;
    }
};
void fun1()
{
    aclass aobj(1);
    try {
        throw("exception in fun1");
        cout <<"print after try in fun1"<< endl;
    }
    catch(const char * s)
    {
        cerr << s << endl;
        throw s;
    }
}
void fun2()
{
    aclass aobj(2);
    try {
        fun1();
        throw("exception in fun2");
        cout <<"print after try in fun2"<< endl;
    }
    catch(const char * s)
    {
        cerr << s << endl;
    }
}
int main()
{
    fun2();
    return 0;
}
```

运行结果：

```
constructing...2
constructing...1
exception in fun1
deconstructing...1
exception in fun1
deconstructing...2
```

【注解】 在 fun2()中调用 fun1()，fun1()的异常从内层抛出到外层，在进行内层抛出的异常处理前，内层函数 fun1()中的局部对象进行析构。

(2) 以 string 类为例，在 string 类的构造函数中使用 new 分配内存，将异常处理机制与其他处理方式对内存分配失败这一异常进行处理对比，说出异常处理机制的优点。

【答】 程序如下:

```cpp
#include <iostream>
#include <new>              //在 Visual C++ 2010 中可以不包含
#include <string>
#include <stdexcept>        //在 Visual C++ 2010 中可以不包含
using namespace std;
class String {
    private:
        char *Str;
        int len;
    public:
        String(int n = 0)
        {
            Str = NULL;
            len = n;
            if(n!= 0)
            {
                try {
                    Str = new char[len];
                }
                catch(bad_alloc& b) {
                    cout <<"throw an exception:"<< b.what()<< endl;
                    throw n;
                }
            }
        }
        ~String()
        {
            if (len > 0)
              delete[] Str;
        }
};
int main()
{
    try
    {
        String s1(100);
    }
    catch(const int n)
    {
        cout << n <<" bytes memory allocate fail!"<< endl;
    }
    return 0;
}
```

【注解】 在类 String 中通过重新抛出异常将构造函数中产生的异常信息抛出(返回);如果采用其他方式,构造函数中的信息将无法传出。

(3) 重载数组下标操作符[],使其具有判断与处理下标越界的功能。

【答】 程序如下:

```cpp
#include <iostream>
using namespace std;
class String
{
private:
    char *Str;
    int len;
public:
    void ShowStr()
    {
        cout<<"string:"<<Str<<",length:"<<len<<endl;
    }
    String(const char *p = NULL)
    {
        if (p)
        {
            len = strlen(p);
            Str = new char[len + 1];
            strcpy(Str,p);
        } else
        {
            len = 0;
            Str = NULL;
        }
    }
    ~String()
    {
        if (Str!= NULL)
            delete[] Str;
    }
    char & operator[](int n)  //重载运算符[],处理 String 对象
    {
        try {
            if(n < 0)
                throw("Underflow exception!");
            if(n >= len)
                throw("Overflow exception!");
            else
                return *(Str + n);
        }
        catch(const char *s)
        {
            cerr << s << endl;
        }
        return *Str;
    }
};
int main()
{
```

```cpp
    String S1("0123456789");
    S1.ShowStr();
    S1[10] = 'A';
    S1[-1] = 'A';
    cout <<" after S1[-1] = A"<< endl;
    S1.ShowStr();
    return 0;
}
```

运行结果：

```
string:0123456789,length:10
Overflow exception!
Underflow exception!
after S1[-1] = A
string:A123456789,length:10
```

【注解】 为了在下标越界产生异常后程序能正常运行，在捕获异常后添加语句"return *Str;"，以便能继续对数组进行存取。

(4) 求负数的平方根、除 0 均为数学类的异常，仿照标准异常处理类，将这类异常用 MathException 类定义，并举一个应用例子。

【答】 程序如下：

```cpp
#include <iostream>
#include <string>
using namespace std;
class MathException
{
private:
    string err_msg;
public:
    MathException(const char *msg):err_msg(msg)
    { }
    const string what()
    {
        return err_msg;
    }
};
float quotient(int a, int b) throw(MathException)
{
    if (b == 0)                //捕获异常
        throw MathException("Divide by 0 !");
    else
        return a/(float)b;
}
int main()
{
    int a, b;
    cout <<"Input a, b: ";
```

```
        cin >> a >> b;
        try
        {
            cout << a <<"/"<< b <<" = "<< quotient(a,b);
        }
        catch(MathException e)
        {
            cerr << e.what()<< endl;
        }
        return 0;
    }
```

运行结果：

```
Input a, b:9 0 ↙
Divide by 0 !
```

第 2 章

C++ 开发环境使用指南

在构思好一个 C++ 程序后,应当在计算机上编辑、编译、调试、运行。现在 C++ 编译器大多数提供了集成开发环境(Integrated Development Environment,IDE),它集编辑、编译、调试、运行于一体,极大方便了编程者。目前 Windows 环境下最为流行的 IDE 为 Microsoft 公司的 Visual C++、Windows 与 Linux 环境下的 CodeBlocks。

2.1 Visual C++ 使用指南

Visual C++ 简称为 VC,是目前使用最广泛的 C++ 开发环境,它是 Microsoft 公司提供的在 Windows 环境下进行 C/C++ 编程的编译器。它不仅是一个集成开发环境,而且是 Microsoft 强大的可视化开发工具集 Visual Studio 中的其中一个。开发 C++ 控制台(console)程序只用到 Visual C++ 的很小一部分功能。

Visual Studio 目前最新版本为 Visual Studio 2019(VS2019),Visual C++ 2010 Express 从 2018 年开始被指定为全国计算机等级考试 C++ 科目使用的软件,本节介绍 Visual C++ 2010 Express 的使用,以下简称为 VC。

2.1.1 启动 Visual C++

启动 VC 有三种方法。

1. 双击桌面图标

Visual C++ 2010 Express 的桌面图标如图 2-1 所示,如果桌面没有图标,可以通过"创建快捷方式"创建。

2. 由"开始"菜单进入

依次启动菜单"开始"→"程序"→Microsoft Visual Studio 2010 Express→Microsoft Visual C++ 2010 Express。

3. 运行 VC 主程序

依次启动菜单"开始"→"运行",在"打开"框中输入"VCExpress"。

VC 启动后,主界面(窗口)如图 2-2 所示。

图 2-1 VC 图标

2.1.2 创建工程

C++ 源程序依赖于工程(project),编辑源程序前先要建立一个工程。

(1) 依次从主界面菜单中选择 File→New→Project,出现如图 2-3 所示的界面。

(2) 选择 Visual C++→Win32 Console Application。

(3) 在 Name 框中输入工程名称"myproject1"。

第 2 章 C++ 开发环境使用指南

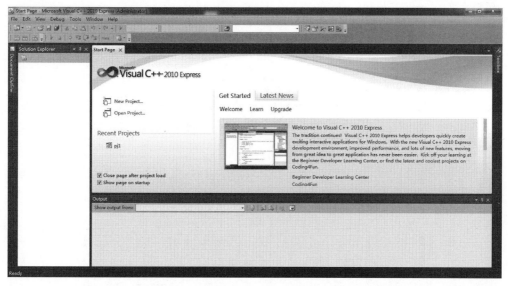

图 2-2 VC 主界面

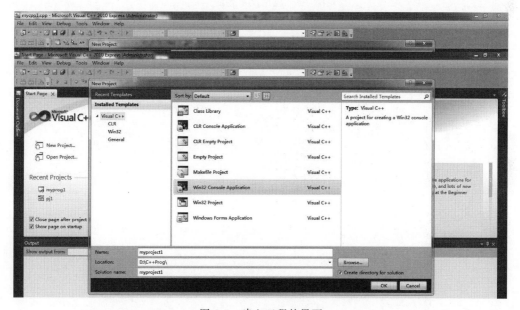

图 2-3 建立工程的界面

(4) 在 Location 框中通过 Browse 按钮指定存放工程的子目录"D:\C++prog"。
(5) 选中 Create directory for solution 复选框,为新工程建立所需目录。
(6) 单击 OK 按钮,出现如图 2-4 所示的应用设定界面。
(7) 单击 Next 按钮,出现如图 2-5 所示的应用设定界面。
(8) 选择应用类型 Application type 选项卡中的 Console application 项。
(9) 选择应用类型 Additional options 选项卡中的 Empty project 项。
(10) 单击 Finish 按钮完成工程设置。
(11) 这时在 D:\C++prog 目录下生成了一个 myproject1 的子目录。

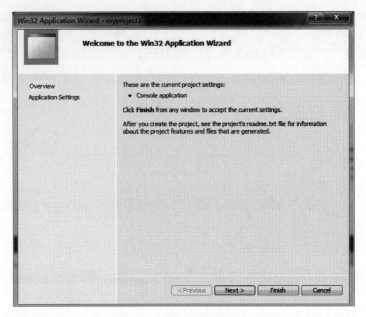

图 2-4 应用设定界面(1)

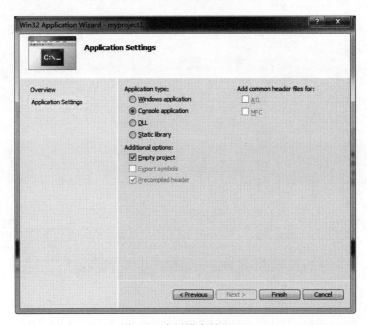

图 2-5 应用设定界面(2)

2.1.3 编辑源程序

1. 建立新的源程序

（1）打开建立的工程 myproject1，如果未出现解决方案资源管理器窗口 Solution Explorer，依次选择 View→Other Windows→Solution Explorer，打开解决方案管理器。

（2）选中工程名或 Source Files，右击，在弹出窗口中选择 Add→New Item，选择过程如图 2-6 所示。

第 2 章　C++开发环境使用指南

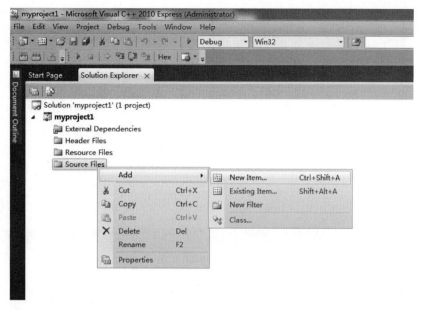

图 2-6　新建文件过程

（3）选择添加项目 Add Item 后，进入如图 2-7 所示的界面。

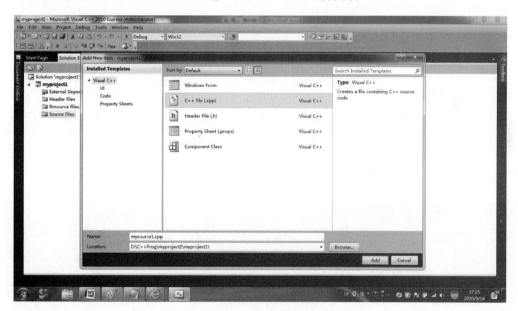

图 2-7　新建源程序

（4）依次选择 Visual C++→C++ File。

（5）在文件名 Name 输入框中输入文件名，如：mysource1。

（6）单击 Browse 按钮，选定存放源文件的子目录，默认目录在工程目录下，如：D:\C++Prog\myproject1\myproject1\。

（7）单击 Add 按钮，进入编辑界面。编辑界面如图 2-8 所示。

（8）这时可以在编辑框中输入源程序了。在输入源程序时，菜单 Edit 中有许多编辑命

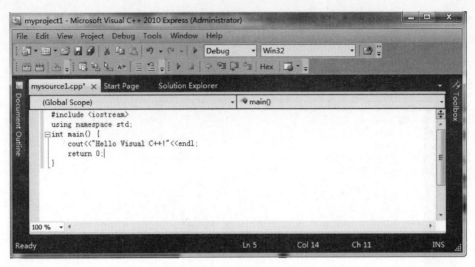

图 2-8 编辑界面

令,如 Copy、Undo 等。在工具栏中也有相应的按钮。除此以外下列热键(快捷键)也是非常有用的:

- Tab 热键:多个空格;
- Ctrl+X 热键:删除块;
- Ctrl+C 热键:复制块;
- Ctrl+V 热键:粘贴块。

程序编辑完成后,可以选择 File→Save 命令保存源文件。保存文件还可以使用快捷键或工具栏中的快捷按钮。

如果想用另外的名字保存文件或将文件存到另外的文件夹中,可以选择 File→Save As 命令。

2. 打开已存在的源程序

选择 Files→Open;或单击工具栏中的"打开"按钮;或使用快捷键打开已保存的程序。

2.1.4 程序的编译与运行

1. 打开与检查工程

工程建立后,打开工程,依次选择 View→Other Windows→Solution Explorer,打开解决方案管理器,检查工程中是否有源程序以及源程序名。

2. 编译生成

选择 Debug→Build Solution,建立解决方案,其中包括编译源程序,如果没有错误,则生成与工程名同名的可执行程序,可执行程序放在工程目录下的 Debug 子目录下。

编译成功与出错信息显示在 Output 窗口,如图 2-9 所示。

也可以单击工具栏中的生成按钮 ![] 或按热键 F7 编译源程序。

3. 运行程序

按 Ctrl+F5 热键运行程序,程序的运行结果在一个 console(控制台)窗口中显示。也可以在 cmd 窗口输入可执行程序名运行可执行程序,如:D:\C++prog\myproject1\debug\myproject1。程序运行界面如图 2-10 所示。

第 2 章　C++ 开发环境使用指南

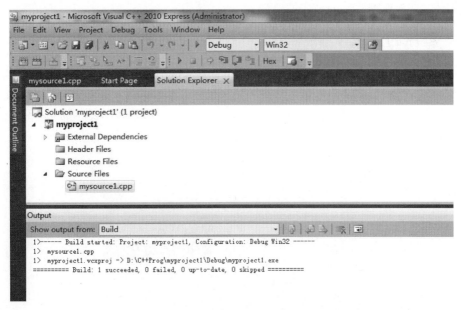

图 2-9　编译生成界面

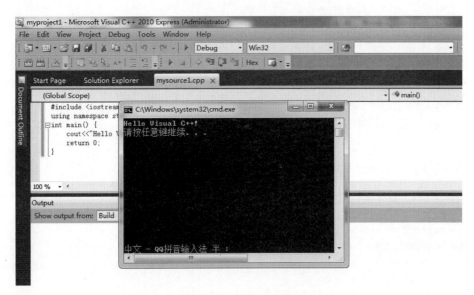

图 2-10　程序运行界面

☆注意：

　　✍ 按 F5 热键或选择 Start Debugging 可能无法运行程序，需要对系统相关项进行设置。

4. 编译与改错

在工程中打开源文件后，就可以选择 Debug→Build Solution，或按 Ctrl＋F7 热键对源文件进行编译。

编译成功或出错的信息均显示在编辑框下的 Output 窗口中，双击一条警告或出错信息，光标就会停在编辑框中相应的行上以便改正。

编译错是一种很容易改正的错误，一般编译器会直接指出出错的行。但是，编译器不会直接指出括号"{}"不匹配、少分号";"的错误。

2.1.5 程序的调试

相对于程序的编译错误,运行错误更难以改正。发现程序中运行错误最好的方法是在程序中加入 cout <<"xx";这样的调试语句显示中间结果。Visual C++编译器提供了非常方便的调试工具 Debug,用于查出程序的运行错误。

1. Debug 的命令

程序编译通过后,在界面工具菜单中选择 Debug,出现如图 2-11 所示的 Debug 命令菜单,工具条上也有相应的调试按钮。

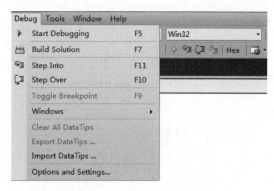

图 2-11　Debug 命令菜单

其中:
- Start Debugging 命令:运行程序;
- Step Into 命令:进入函数体、循环体等逐条运行;
- Run to Cursor 命令:控制程序运行到编辑框的光标指示的源程序处。

下列程序进行 m~n 的累加:m+1+m+2+…+n:

```
1    #include<iostream>
2    using namespace std;
3    int Sum(int m,int n = 1) {
4        int t,s = 0;
5        if(m > n) { t = m; m = n; n = t;}
6        for(int i = m; i <= n; i++)
7            s = s + i;
8        return s;
9    }
10   void main()
11   {
12       int m,n,s;
13       cin >> m >> n;
14       s = Sum(m,n);
15       cout <<"Sum("<< m <<","<< n <<") = "<< s << endl;
16   }
```

现将光标停在输入字符串的语句"cin >> m >> n;"上,启动 Run to Cursor 命令,程序运行到 cin >> m >> n;处停下,这时在编辑框下面出现 locals 与 Breakpoints 框,如图 2-12 所示。

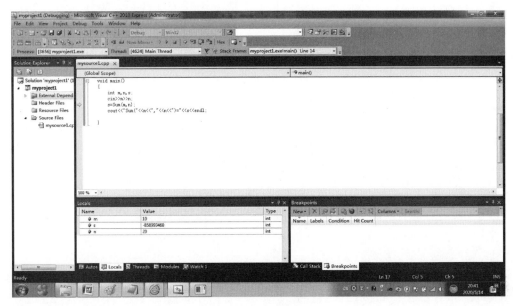

图 2-12　调试界面

2. Locals 与 Watch 窗口

Locals 窗口按函数显示各变量的值,例如,图 2-12 中显示了 main()函数中 m、n、s 的值。watch 窗口可显示 name 框中输入的表达式 m+n 的值。

3. debug 命令的配合使用

debug 启动后程序运行到 cin >> m >> n;处停下后,激活了 Debug 的其他命令:

- Stop Debugging:结束 Debug;
- Step Out:退出函数体;
- Step Over:执行指向的函数、语句。

运行到 cin >> m >> n;处停下后,选择 Step Over 命令,执行此语句等待输入。

☆**注意**:

如果此时选择 Step Into 命令,程序就会陷入 cin 的代码中。

输入数据后,程序在"s=Sum(m,n);"处停下,此语句中带有函数,选择 Step Into 命令进入 Sum(m,n)函数,程序在 **int Sum(int m,int n=1)** 处停下(如果选择 Step Over 则整个语句连同函数就执行了)。

再选择 Step Into 命令进入函数体,在"int t,s=0;"处停下,选择 Step Over 命令执行此语句,在 for(int i=m;i<=n;i++)前选择 Step Into 命令进入循环。这时可以在 watch 窗口中输入变量 i,可以观察变量 i 随程序运行值的改变,并可在 watch 中人为改变 i 的值。

选择 Step Out 退出函数 **int Sum(int m,int n=1)**,程序在"s=Sum(m,n);"处停下,选择 Step Over 执行此语句。最后选择 Stop Debugging 退出 Debug。

2.1.6　多文档工程

一个工程可由多个程序组成,可以将上述程序的 int Sum(int m,int n=1)函数放到

mycpp1_1.cpp 中，编译时加入工程 myprog，生成 mycpp1_1.obj。

将 void main()函数体放到 mycpp1_2.cpp 中，编译时由于 Sum()函数没有在 main()中定义，此时编译出错，因此要在 mycpp1_2.cpp 中进行 Sum()函数原型声明。编译时 mycpp1_2.cpp 加入工程 myprog 中，生成 mycpp1_2.obj。

由 mycpp.cpp 分成的两个程序如下：

```
                                          mycpp1_2.cpp
    mycpp1_1.cpp                          # include <iostream>
1   # include <iostream>                  using namespace std;
2   using namespace std;                  int Sum(int m,int n = 1);
3   int Sum(int m,int n = 1) {            void main()
4       int t,s = 0;                      {
5       if(m > n){t = m;m = n;n = t;}         int m,n,s;
6       for(int i = m;i <= n; i++)            cin >> m >> n;
7           s = s + i;                        s = Sum(m,n);
8       return s;                             cout <<"Sum("<< m <<","<< n <<") = "<< s << endl;
9   }                                     }
```

当试图生成 myprog.exe 时出错，因为在工程 myprog 中有 3 个源程序，其中 mycpp1_2.cpp 与 mycpp.cpp 中都含有 main()函数。启动 view→Solution Explorer，打开 Source Files 选项卡，选中源程序文件中的 mycpp.cpp，按 Delete 键在工程中删除 mycpp.cpp。

☆注意：

☞若一个工程包含多个源程序，必须只有一个源程序含有 main()函数。

2.2 CodeBlocks 使用指南

Code::Blocks(CodeBlocks，简称为 CB)是一个开源、免费、跨平台的 C++集成开发环境。它基于著名的图形界面库 wxWidgets，用 C++编写开发，并捆绑了 MinGW 编译器。目前 Windows 与 Linux 操作系统都有 CodeBlocks。

2.2.1 CodeBlocks 的安装与配置

CodeBlocks 的官方网站为 http://www.codeblocks.org，最新版本为 13.12，有 Windows 和 Linux 版，Windows 版本的文件有：

codeblocks-13.12-setup.exe

codeblocks-13.12mingw-setup.exe

前一个文件不带编译器，后一个文件带有 MinGW 的 GCC 编译器和 GDB 调试器。

1. 安装

如果想使用 VC 编译器，下载第一个文件；如果想使用 GCC 编译器，下载后一个文件，双击安装。

安装过程为：

(1) 对 Choose Components 选择全部安装"Full：All plugins，all tools，just everything"。

(2) 对 Choose Install Location 选择默认文件夹 C:\Program Files(x86)\CodeBlocks。

(3) 安装完毕选择 run code::blocks now。

(4) 进入 CodeBlocks 启动画面,如图 2-13 所示。

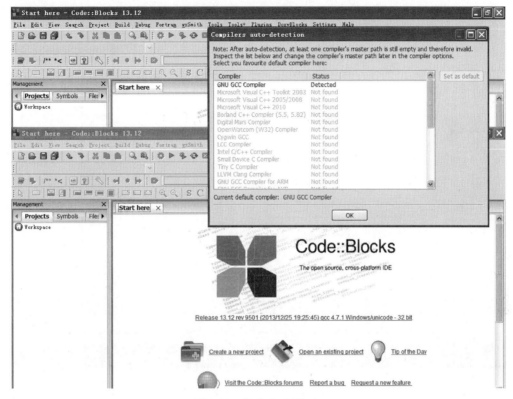

图 2-13　启动 CodeBlocks

2. 设置 CodeBlocks

第一次启动 CodeBlocks 会检查计算机中的编译器,出现图 2-13 中的 Compilers autodetection 对话框,其中显示自动检测到 GNU GCC Compiler 编译器,对话框右侧的 Set as default 按钮,显示将它设成默认的编译器,单击 OK 按钮,确认将 GCC 设成默认的编译器。

如果不对 CodeBlocks 进行特殊的设置,此时可以进入编程了。

若要对 CodeBlocks 进行设置,选择 Settings 菜单项,显示需要进行设置的项目,设置项目如图 2-14 所示。

设置项目有 Environment(环境)、Editor(编辑器)、Compiler(编译器)、Debugger(调试器)等。

1) 环境设置

选中 Environment 项,显示一个环境设置窗口,窗口的左边显示设置的图标与项目名,环境设置的项目有 Help Files(帮助文件)、Global compiler settings(编译器设置)、Autosave(自动保存设置)等。环境设置窗口如图 2-15 所示。

图 2-14　设置菜单项

CodeBlock 只提供自身如何使用的帮助,而 C++ 语言本身的帮助需要自己添加到系统中,以便随时查阅。CodeBlocks 允许将 MSDN 之类的帮助文档添加到系统中,添加帮助文档的步骤如下:

(1) 下载帮助文档,如 cppreference.htm,并将它放到 CodeBlocks 目录下。

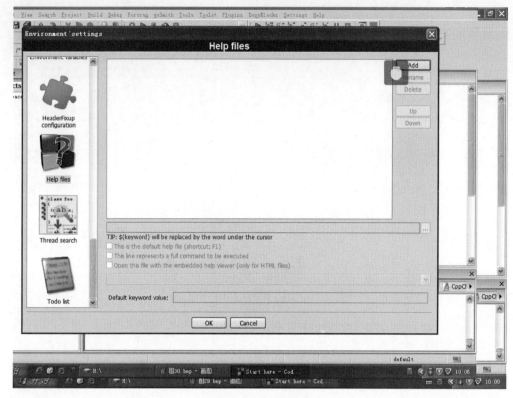

图 2-15 环境设置窗口

（2）单击 Environment 窗口左边的 Help Files 图标，显示图 2-15 所示的窗口。

（3）单击右上角 Add 按钮，弹出 Add title 对话框。

（4）在对话框中输入帮助文件的标题，如 C++Manual，单击 OK 按钮。

（5）单击 ... 按钮，进入到帮助文件的文件夹，将帮助文件 cppreference.htm 加入。

（6）为了方便使用，选中 This is the default help file（shortcut：F1）复选框，然后单击下面的 OK 按钮。

（7）选择主菜单的 Help，发现子菜单项中含有 C++Manual F1，表明文档添加成功。选中此项，或按 F1 热键，进入帮助窗口。

为防止编写或者调试程序的过程中偶尔出现断电，造成部分程序内容丢失，可以分别设置 Autosave 中的源程序与项目文件的自动保存时间。

2）编辑器设置

编辑器的设置项目有通用设置 General settings、源代码格式设置 Source formatter 等。

- General settings 可以设置编辑器的字体，文字大小等；
- Source formatter 用来设置源代码的格式，CodeBlocks 提供了 Allman(ANSI)、K&R、Linux、GNU、Java 等风格，并可以自定义风格。

2.2.2 编辑源程序

双击 CodeBlocks 图标，进入 CodeBlocks 界面。

1. 建立新工程

源程序要依附于工程(project)，依次选择编译器主界面工具菜单中的 File→New→Project，出现如图 2-16 所示的工程类型选择界面。

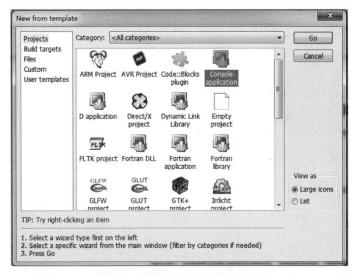

图 2-16　工程类型选择界面

选择 Console application 图标，单击 Go 按钮后，进入语言类型选择窗口，窗口中有 C、C++语言供选择。选择 C++，单击 Next 按钮，进入工程名填写窗口，如图 2-17 所示。

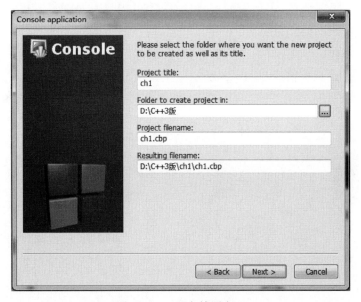

图 2-17　工程名填写窗口

在 Project title 框中输入工程名，如 ch1，在 Folder 框中指定工程存放的文件夹(位置)，第 3 个框中自动按工程名生成工程文件名 ch1.cbp，第 4 个框中显示工程文件名的路径及文件名。单击 Next 按钮，进入图 2-18 所示的编译器选择窗口。

在 Compiler 下拉框中有很多编译器可选，如 GNU GCC Compiler、Microsoft Visual C++ 2010、

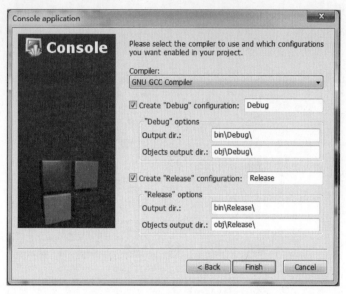

图 2-18　编译器选择窗口

Borland C++ Compiler(5.5,5.82)等,由于 CodeBlocks 安装时只装入了 GCC Compiler,因此,选择 GNU GCC Compiler。

下面的两个复选框中选择是否生成调试版 Debug 和发行版 Release,这两个选项要求至少选一项,一般在程序调试期间或编写短程序时选择调试版;发行版是在程序调试好后,将其编译成发行的软件时的选项。在编译发行版时要进行优化,占用时间,同时发行版文件占用空间,因此,平时不选择生成发行版。

2. 建立新的源程序

建立工程后,就可以编辑源程序了。

依次选择主界面的 File→New→File,出现如图 2-19 所示的文件类型选择界面。

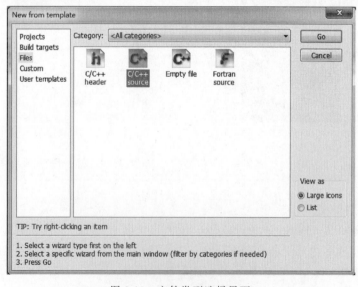

图 2-19　文件类型选择界面

选择 C/C++ 图标，单击 Go 按钮，在 C 与 C++ 中选择 C++，单击 Next 按钮，进入图 2-20 所示的文件名输入窗口。

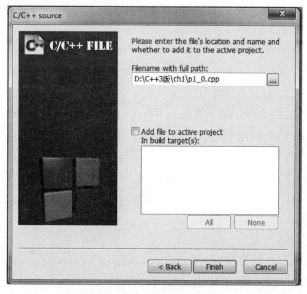

图 2-20　文件名输入窗口

单击 按钮，选择存放源程序的文件夹，并输入带 .cpp 扩展名的文件名，如 p1_0.cpp，选中 Add file to active project 复选框，单击 Finish 按钮，进入主界面，光标停在源程序编辑区。编辑界面如图 2-21 所示。

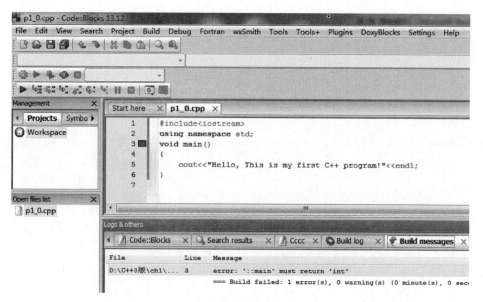

图 2-21　编辑界面

这时可以在编辑框中输入源程序了，在输入源程序时，菜单 Edit 中有许多编辑命令，如 Copy、Undo 等。在工具栏中也有相应的编辑按钮。除此以外，2.1.3 节介绍的热键（快捷键）也是非常有用的。

当程序编辑完成后,可以选择 File→Save 或单击工具栏中的"保存"按钮保存源程序。

3. 打开已存在的源程序

选择 File→Open 或单击工具栏中的"打开"按钮可打开已保存的源程序。

☆注意:

- CodeBlocks 允许不建立工程直接编辑与打开 C++ 程序并编译运行,生成的可执行文件与源程序文件在一起。
- 直接建立文件的方法是依次选择 File→New→Empty file,或单击 New file 图标,或按热键 Ctrl+Shift+N,这时建立一个名为 Untitledx 的编辑区,当需要保存文件或要编译时,需要给出文件名(包括位置)。
- 直接建立或打开的程序不能用 Debugger 进行调试。当不需要用调试器进行调试时,为简便起见,可直接建立与打开源程序文件。

2.2.3 程序的编译与运行

1. 程序的编译与修改

当源程序编辑完成后或打开后,选择 Build→Compile current file,或使用热键 Ctrl+Shift+F9 对程序进行编译。如果出现一个信息窗口显示下列信息:

```
That file isn't assigned to any target
```

说明文件还没加入到任何目标中,需要进行下列操作:

选择 File→Add Files,进入源文件所在的文件夹,将 p1_0.cpp 加入工程中。

程序的编译信息显示在图 2-21 所示的 Logs&others 窗口中的 Build messages 栏中,编译信息包括警告与错误信息。

如果编译通过(没有错误),就会在源程序的文件夹中生成带 .obj 后缀的目标文件。例如 p1_0.cpp 的目标文件为 p1_0.obj。

如果编译未成功(有错误),Build messages 窗口中将列出出错行号及出错信息,编辑窗口程序的行号前会有一个红"口"标识第一个出错的行。双击信息窗口的某条出错信息,光标就会停在编辑框中出错的行上,以便改正。

图 2-21 中显示第 3 行":: 'main' must return 'int'",错误原因是 main()函数必须返回 int,将"void"改成"int",main()函数最后一行加上"return 0"后编译成功。

☆注意:

- 程序编译的信息显示在 Logs 窗口中,要在主菜单的 View 中打开 Logs 窗口,否则将看不到任何信息。可以按热键 F2 打开/关闭 Logs 窗口,热键 F1 常用于帮助,热键 F2 用来打开信息窗口,可见其重要。

2. 生成与运行

编译成功只说明源程序没有语法错误,要运行程序先要生成(制作)可执行程序:

选择 Build→Build ctrl+F9,或单击工具菜单的 build 图标,或按热键 Ctrl+F9。

此时如果 Build messages 框中出现下列错误信息:

```
multiple definition of 'main'
```

表明工程中含有多个 main()函数,将主窗口左边的 Project 栏中 Workspace 下的 sources 展开,发现其中还有 main.cpp 这个程序,这是系统生成的一个程序,其中含有 main() 函数。制作执行文件不能含有两个 main(),此时需要在工程中移除 main.cpp。移除 main.cpp 的方法如下:

选中 main.cpp,右击,在弹出的快捷菜单中选择 Remove file from project,或者选中 main.cpp,选择主菜单 Project→Remove files。

此时可能会出现图 2-22 所示的消息框,显示工程的可执行文件没有建立。其原因是源程序 p1_0.cpp 没有加入到工程 ch1 中。

单击"是(Y)"按钮,将建立可执行程序。

当执行文件建立后,可以选择 Build→Run,或单击工具栏中的"运行"按钮 ▶,或使用热键 Ctrl+F10 运行可执行文件。

还可以选择 Build→Build and run,或使用热键 F9 生成并运行可执行文件。

这时弹出结果显示窗口,如图 2-23 所示。

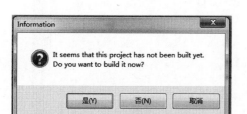

图 2-22 工程未建立消息框

图 2-23 结果显示窗口

可以对结果显示窗口的信息进行操作,方法是:右击,在弹出的快捷菜单中按菜单项的提示进行操作。

☆注意:

☑ 结果显示窗口查看完毕后要关闭,否则 CodeBlocks 不能进行新的编译。

☑ 可执行程序名与工程名相同,存放在工程文件夹下的"\bin\debug"下。Run 运行程序运行可执行文件;程序修改后必须重新 Build,否则运行的是上次生成的可执行程序。

3. 编辑工程

工程的主要内容为源程序文件,可以在工程中添加源程序文件、删除源程序文件。

添加源程序文件的步骤为:选择 File→Add Files,进入源文件所在的文件夹,双击源程序将源程序加入到工程中。

删除源程序文件的步骤为:选中 Workspace 中 source 列表中的源程序,右击,在弹出的快捷菜单中选择 Remove file from project,或者选中源程序,选择主菜单 Project→Remove files。

选择 File→Save project 或 Save project As 项,或单击工具栏中的 图标将工程保存。

如果以前生成了工程,可选择 File→Open,在工程文件所在的文件夹中打开扩展名为 .cbp 的文件。

2.2.4 查帮助

系统的帮助文档对程序员来说至关重要,从帮助文档中,程序员可以获得很多从课本中难以获得的东西。CodeBlocks 中自带如何使用集成环境的帮助文档,编译器的帮助文档需要在设置系统时自己安装。

经常要查的文档是函数原型与功能,查函数功能有两种方法:按目录查与搜索关键字。

1. 查函数分类目录

如果不知道函数名,可查函数目录,C++ 的函数在帮助文档中分类列出。按下列顺序进行。

在主界面主菜单中选择 Help→C++ Manual(C++ Manual 为载入帮助文档时取的菜单名),出现帮助文档窗口,如图 2-24 所示。

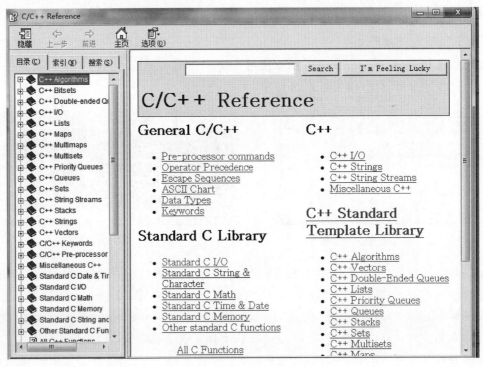

图 2-24 C/C++ 帮助文档窗口

在"目录"选项卡中依次展开所含内容,或在 C/C++ Refrence 区域选择主题,列出该主题的函数列表。

2. 搜索函数

如果知道函数全名,在"搜索"框中输入函数名进行查找;若函数全名记不清,选择"索引"选项卡,在"键入查找单词的头几个字母中"对话框中输入函数名,双击列出的函数名,右边的框中显示该函数的详细功能与实例。

2.2.5 程序的调试

相对于程序的编译错误,运行错误更难以改正。发现程序中运行错误最好的方法是在程序中加入 cout<<"xx"这样的调试语句以显示中间结果。CodeBlocks 编译器提供了非常方便的调试工具 Debugger 用于查出程序的运行错误。调试程序之前,首先要确认已经设置了 Produce debugging symbols [-g] 选项,按照如下顺序去找:

Project→Build options…→Compiler settings→Debug compiler Flags

如果 Prodece debugging symbers [-g] 项已经配置,则可以调试程序了,否则选中该复选框,单击 OK 按钮。

1. Debug 命令与使用

Debug 命令在主菜单 Debug 中,如图 2-25 所示。

除了在 Debug 菜单中使用 Debug 命令外,常用的命令在工具图标中列出,并设置了热键。从图 2-25 中看出,不是所有 Debug 命令都可用,哪个命令是否可用依赖于一定的状态。

这些命令的功能使用方法如下:

首先,要在程序中设置一个点,让程序运行到这个位置暂停,以便检查此刻的运行结果;把光标置此行代码前,再用 Debug→Run to cursor 或者调试工具菜单中的 Run to Cursor 按钮或热键 F4。此时程序运行到此处就停下来,此处称为断点。也可以用 Debug→Toggle breakpoint 设置断点,然后选择 Debug→Start/Continue 来启动调试器。

若在一个已设置断点前再次使用 Toggle breakpoint,则该断点就被删除,如果想删除所有断点,可以用 Debug→Remove all breakpoints。

如果希望运行到程序下一行前面,则可以选择 Debug→Next line。如果不想每次执行一行代码,希望执行的单位更小,可以用逐条执行指令(Next instruction)。如果希望运行到某个代码块(例如,调用某个函数),可以选择菜单 Step into,则运行到该代码块内。如果希望跳出该代码块,则可以选择菜单 Step

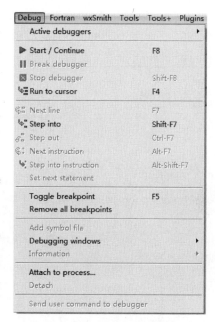

图 2-25　Debug 命令菜单

out。如果希望终止调试器,可以选择菜单 Start/Continue,则调试器会自然运行到结束,也可以用 Stop debugger,则调试器就会被强行终止,调试过程结束。

如果希望查看程序中变量的值,可以选择 Debug→Debugging windows 子菜单的 Watches 项。

2. 使用实例

下列程序的功能实现将一个输入的字符串颠倒后显示。

```
1   #include<iostream>
2   #include<string.h>
3   using namespace std;
4   char * reverse(char * str) {
5       char t;
6       int i;
7       for(i=0; i<strlen(str); ++i) {
8           t=str[i];
9           str[i]=str[strlen(str)-i];
10          str[strlen(str)-i]=t;
11      }
```

```
12      return str;
13  }
14  int main()
15  {
16      char s[180];
17      cout <<"Enter a string:";
18      cin >> s;
19      //cout <<"strlen:"<< strlen(s)<< endl;
20      cout <<"Reverse string:"<< reverse(s)<< endl;
21      return 0;
```

运行结果：

```
Enter a string:abcde fg
Reverse string:abcde

Process returned 0 (0x0)    execution time : 12.527 s
Press any key to continue.
```

结果显然与期望的不一致，初步判断错误出在 reverse() 函数中，下面看看 reverse() 执行的详细过程。

☆**注意**：

- CodeBlocks 工程路径中包含汉字可能导致断点不停。因此，如果工程路径包含汉字，应将其移到不包含汉字的路径下。

1）设置断点

将光标置于第 14 行 int main() 前，单击 Run to cursor 按钮或按 F4 热键，可以看到第 17 行前面有个箭头，表明程序运行到这里停下了，而且还出现了一个没有任何输出信息的窗口。

或将光标置于第 14 行 int main() 前，选择 Toggle breakpoint 菜单项或按 F5 热键，这时第 14 行前出现圆点●标识该断点；选择 Start/Continue 菜单（红三角按钮）或按 F8 热键，程序同样运行到第 17 行停下。

2）观察执行

为了查看程序运行中变量值的变化情况，需要打开观察变量的窗口（单击 Watches 按钮，或选择菜单 Debug→Debugging windows→Watches）。为了方便观察整个调试过程布局，拖动 Watches 窗口到右边，并展开各个变量。从 Watchs 窗口中看到此时局部变量 s 的值为空，如图 2-26 所示。

选择 Next Line 或按热键 F7，执行程序第 17 行 cout…，显示输入提示信息，此时结果窗口会出现提示输入信息"Enter a string:"。

继续按热键 F7，执行第 18 行 cin…，此时，黄色箭头不见了，需要在结果窗口中输入字符串，仍输入"abcde fg"。回到调试窗口发现 watchs 窗口的 s 变量的值变成了"abcde"。第 19 行是注释行，黄箭头指向第 20 行。观察窗口如图 2-27 所示。

第 20 行调用 reverse() 函数，返回处理后的字符串，如果继续按热键 F7 将执行该行，则越

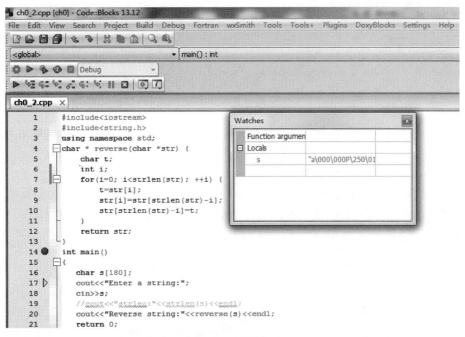

图 2-26 调试界面

过了 reverse()函数,观察不到函数的执行过程。这时需要进到函数体内。单击 Step into 按钮,或选择菜单 Debug→Step into,或按 Shift+F7 热键,进入 reverse()函数。执行点越过局部变量定义,指向第 9 行 for…,观察窗口如图 2-28 所示。

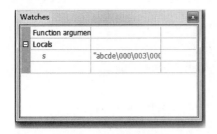

图 2-27 观察窗口(1)

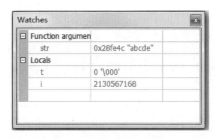

图 2-28 观察窗口(2)

☆**注意**:

✎ 如果此时选择 Next line 命令,程序就会越过函数。

按热键 F7 进入 for 循环中,这时为了观察 strlen(str)的值,需要在观察窗口中加入 strlen(str),加入后,显示其值为 5,表明正确计算了字符串的长度,如图 2-29 所示。

继续按热键 F7,执行完第 8 行,t 改成'a'。当一个变量的值变化时,观察窗口中该项变成红色。

继续按热键 F7,执行完第 9 行,发现 str 变空了,strlen(str)变成 0 了。原来执行了 str[0]='\0',将 str 第一个字符置成了结尾符'\0'。其原因是 str[strlen(str)-i]指向了串尾的结束符(应该将它改成 str[strlen(str)-i-1])。

错误找到后,选择 Step out 或按 Ctrl+F7 热键跳出函

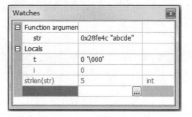

图 2-29 加入了函数的观察窗口

数,执行点指向第 20 行。

选择 Debuger→Stop debugger 或按 Shift+F8 热键结束调试。

现在修改程序,将第 9、10 行的 str[strlen(str)-i]改成 str[strlen(str)-i-1]。重新运行程序,发现结果竟然还是一样。如何发现错误,留给读者练习!

2.2.6 多文档工程

一个工程可由多个程序组成,可以将上述程序的 char * reverse(char *)函数放到源程序 mybcpp1_1.cpp 中保存。然后,启动 Project→Add files,将它加入工程 myprog。

将 void main()函数体放到源程序 mybcpp1_2.cpp 中,编译时由于 reverse()函数没有在 main()中定义,此时编译出错,因此要在 mybcpp1_2.cpp 中进行 reverse()函数原型声明。

启动 Project→Add files,将 mycpp1_2.cpp 加入工程 myprog 中,并启动 Project→Remove files,将工程中原有的文件 mybcpp.cpp 从工程 myprog 中撤销。

由 mybcpp.cpp 分成的两个程序如下:

mybcpp1_1.cpp
```
1  #include<string.h>
2  char * reverse(char * str) {
3    char t;
4    int i;
5    for(i=0;i<strlen(str)/2;++i){
6      t = str[i];
7      str[i] = str[strlen(str)-i-1];
8      str[strlen(str)-i-1] = t;
9    }
10   return str;
11 }
```

mybcpp1_2.cpp
```
#include<iostream>
char * reverse(char *);
using namespace std;
int main(){
  char s[180];
  cout<<"Enter a string:";
  cin>>s;
  cout<<"Reverse:"<<reverse(s)<<endl;
  return 0;
}
```

启动 Project 菜单并选择 Build Project,重新生成 myprog.exe。

☆注意:

☑ 若一个工程包含多个源程序,必须只有一个源程序含有 main()函数。

2.3 Linux GNU g++ 上机编程指南

Linux 环境中有许多 IDE,如 Glade、Kdevelop、Anjuta、Kylix 等,但它们没有附带在 Linux 系统盘中,Linux 系统中附带的 C++开发工具为 EMACS、g++、gdb 的组合,它们深受广大程序员的喜爱。

2.3.1 使用 EMACS 编辑源程序

EMACS 为 Editor MACros 的缩写,是一个以编辑器为主干的环境软件,它不仅可以用来编辑文本,在编辑器的基础上提供了一个整合的工作环境,还可以在 EMACS 环境中完成如发电子邮件、浏览 Web、编译程序等工作。

1. EMACS 的启动与退出

在文本模式下,输入 EMACS 即可启动 EMACS:

[root@localhost yang]# emacs ⏎

启动后的 EMACS 做了如下的初始化(initialize)工作。

(1) 清除目前的屏幕,开始一个全新的 EMACS 屏幕。

(2) 显示 EMACS 的基本信息。其中包括:目前使用的 EMACS 版本、基本的帮助信息以及有关 EMACS 版权信息等。此时若不输入任何指令,EMACS 会在一段时间后(约 2 分钟)自动将屏幕重新清除成一个空白的屏幕。

(3) 若在屏幕自动重新清除之前输入命令,EMACS 会执行命令进行相应的操作。

EMACS 启动后,窗口构成如图 2-30 所示。

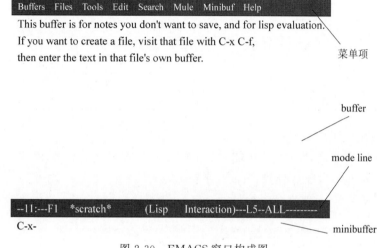

图 2-30　EMACS 窗口构成图

其中:
- "菜单项"显示了 EMACS 的菜单功能分类;
- buffer 是操作缓冲区,分为文件编辑缓冲区与信息显示缓冲区。文件编辑缓冲区用于文件编辑,信息显示缓冲区用于显示帮助信息、菜单项的信息等;
- mode line 是操作缓冲区的状态栏,显示文件名(或缓冲区名)、光标所在行号等;
- minibuffer 为命令行,显示输入的命令以及用来输入子命令。

EMACS 启动后的缓冲区仅用于显示信息,在其中输入的信息不能保存。

按 Ctrl+Z 热键暂时离开 EMACS 回到其上一层的状态,一般是回到 shell 的状态。若想回到 EMACS 的状态,只要输入%emacs。

按 Ctrl+X Ctrl+C 热键也可退出。按 Ctrl+X Ctrl+C 热键的方法为按 Ctrl+X 热键后再按 Ctrl+C 热键,也可以按住 Ctrl 键不放连续按 X、C 键。

按 EMACS 键同时输入文件名可以直接进入文件编辑:

[root@localhost yang]# emacs　mycpp.cpp ⏎

此时,缓冲区(窗口)为文件编辑缓冲区,在窗口中可以输入文本。

在 X-Windows 下可以通过启动"主菜单"→"编程"→Emacs,进入 EMACS。

2. EMACS 的命令格式与常用命令

EMACS 大部分操作是通过快捷键来实现的,当然也可以通过选择菜单实现,但使用快捷

键比使用菜单方便得多。EMACS 命令(快捷键)通常使用下列组合键:

Ctrl 键:如 Ctrl+X 和 Ctrl+C 热键,由于此命令在命令行显示为"C-X-C",因此简记为"C-X C-C";

Alt 键:如 Alt+X 热键,表示为"Meta-X"在命令行显示为"M-X";

Esc 键:若键盘上无 Alt(Meta)键,可用 Esc 键代替,"Esc-X"表示按了一个 Esc 键后再按一个 X 键。

表 2-1 列出了 EMACS 的常用命令。

表 2-1 常用命令

命　　令	功　　能	说　　明
Ctrl+H	帮助	输入此命令后,可输入?或字母进一步选择帮助子项
Ctrl+H　T	进入 EMACS 使用说明	字母 t 为帮助子项"tutorial"的头字母
F10	启动菜单	按上、下箭头翻看菜单栏目,按 PgUp 键选择菜单子项
Ctrl+G	撤销输入的命令	例如,可以终止重复输入的搜索命令 Ctrl+s
Ctrl+X Ctrl+C	退出 EMACS	

上述命令使用说明如下。

(1) 按 Ctrl+H 热键启动帮助功能,然后输入一个字符子命令来说明需要哪种帮助。Ctrl+H T 表示按 Ctrl+H 热键,然后输入字符 t。

(2) F10 启动菜单,这时在命令行显示:

Menu bar(up/down to change, PgUp to menu):b==> Buffers

同时开辟一个菜单显示缓冲区窗口。

按上、下键翻阅菜单项,按 Enter 键选中;或直接输入一个字符子命令来选中菜单项。

选中菜单项后,命令行显示菜单子项供选择。同样,按上、下键翻阅菜单项,按 Enter 键选中;或直接输入一个字符子命令来选中菜单子项。

也可以按 PgUp 键进入显示缓冲区窗口选择菜单项与子项。

选中菜单子项后就会执行相应的功能。

(3) 如果打开了一个编辑缓冲区,在执行 Ctrl+X、Ctrl+C 命令退出 EMACS 前,EMACS 会再三地提醒使用者有关文件存档与仍在执行的程序的信息。命令行会提示(y, n, !, ., q, Ctrl+R or Ctrl+H)让使用者选择。

- y:对命令行所显示的缓冲区存档,并询问对其他缓冲区是否存档的意见。
- n:放弃对命令行所显示的缓冲区存档,并询问对其他缓冲区是否存档的意见。
- !:同意对命令行所显示的缓冲区存档,且对其他的缓冲区也一并存档。
- .:同意对命令行所显示的缓冲区存档,直接放弃其他缓冲区的存档。
- q:放弃存档状态而不执行任何存档的动作。

3. 基本编辑命令

EMACS 编辑框中可以使用 PgUp、PgDn、←、↑、→、↓ 移动光标,与之等功能的移动光标命令不做介绍。基本编辑命令如表 2-2 所示。

4. 文件与缓冲区(窗口)命令

文件与缓冲区(窗口)命令如表 2-3 所示。

表 2-2　基本编辑命令

功能分类	命令	功能
光标移动	Ctrl+A	将光标移动到行首
	Ctrl+E	将光标移动到行尾
字符操作	Delete	删除光标处的字符
	Backspace	删除光标前的字符
行操作	Ctrl+K	删除(剪切)从光标处到行尾的字符
	Enter	换新行
块操作	Ctrl+Space	设置块头
	Alt+W	设置块尾
	Ctrl+Y	粘贴设定块的内容或按 Ctrl+K 键删除的内容
查找与替换	Ctrl+S	从当前位置向尾找,按 Ctrl+S 键继续向后找
	Ctrl+R	从当前位置向前找,按 Ctrl+R 键继续向前找
	Alt+X	启动查找与替换功能,在命令 M-x 后继续输入 replace,query 子命令分别进行全部与选择性替换
撤销	Ctrl+X U	撤销编辑,重复使用 Ctrl+X U 可撤销多次操作,最多撤销 20 次

表 2-3　文件与缓冲区(窗口)命令

功能分类	命令	功能	说明
文件操作	Ctrl+X Ctrl+F	在缓冲区中打开文件	在命令行输入文件名
	Ctrl+X Ctrl+S	将缓冲区的内容保存到文件	不关闭缓冲区
	Ctrl+X Ctrl+W	将缓冲区的内容保存到指定的文件	在命令行输入文件名
缓冲区(窗口)操作	Ctrl+X Ctrl+B	列缓冲区(窗口)	可进入窗口列表按 Enter 键选择窗口
	Ctrl+X o	切换窗口	o 为英文字母,非数字 0
	Ctrl+X 1	关闭其他窗口,只保留当前窗口	1 为数字,关闭的窗口并未消失
	Ctrl+X 2	将屏幕水平分割成两个窗口	两个窗口的内容完全一致
	Ctrl+X 3	将屏幕垂直分割成两个窗口	

5. 其他命令

运行 shell 命令如表 2-4 所示。

表 2-4　其他命令

命令	功能	说明
Esc!	运行 shell 命令	运行的结果显示在当前缓冲区中

例如在使用 EMACS 时,要看文件目录,在输入"Esc!"命令后,在命令行输入:

　　Shell command:dir -l↙

将当前目录下的文件列到窗口中。

EMACS 还有许多命令,上述只列出了很少的一部分以满足基本的需要,如果还需要使用更多的命令可以查看帮助,也可以通过菜单进行操作。在 Linux 提供的窗口环境 X Window 下,EMACS 的菜单操作更方便。

6. 命令使用举例

首先输入一个程序,程序名为 mycpp.cpp,其功能为计算 $m \sim n$ 的累加 $m+1+m+2+\cdots+n$。

[root@localhost yang]# emacs mycpp.cpp ↙

启动窗口如图 2-31 所示。

```
Buffers   Files   Tools   Edit   Search   Mule   C++   Help
#include<iostream>
using namespace std;
int Sum(int m,int n=1) {
    int t,s=0;
    if(m>n) { t=m; m=n; n=t;}
    for(int i=m; i<=n; i++)
        s=s+i;
    return s;
}
void main()
{
    int m,n,s;
    cin>>m>>n;
    s=Sum(m,n);
    cout<<"Sum("<<m<<","<<n<<") = "<<s<<endl;
}
--11:---F1 myprog.cpp        (C++)----L1---ALL--------------
```

图 2-31 编辑窗口

在窗口中输入完程序,再输入如下命令将程序保存:

Ctrl+X Ctrl+S

如果要退出编辑,按 Ctrl+X Ctrl+C 热键。

现在要将 mycpp.cpp 分成两个文件,一个为 mycpp_1.cpp,另一个为 mycpp_2.cpp。分别存放 Sum()函数与 main()函数。

(1) 按 Ctrl+X 2 热键将屏幕分成两个窗口。

(2) 按 Ctrl+X Ctrl+F 热键,输入文件名 mycpp_1.cpp,即可在 mycpp_1.cpp 窗口中编辑此文件。

(3) 按 Ctrl+X O 热键将窗口切换到 mycpp.cpp。

(4) 将光标指到文件头,按 Ctrl+Space 热键设定块首。

(5) 将光标指到函数尾,按 Alt+W 热键设定块尾。

(6) 按 Ctrl+X O 热键将窗口切换到 mycpp_1.cpp。

(7) 按 Ctrl+Y 热键将块的内容粘贴到 mycpp_1.cpp。

(8) 按 Ctrl+X Ctrl+S 热键保存 mycpp_1.cpp。

(9) 使用同样的方法开辟 mycpp_2.cpp 窗口,把 main()函数搬入。

(10) 最后按 Ctrl+X Ctrl+C 热键退出。

☆注意:

✎ 要分清编辑窗口与信息显示窗口,不要在信息显示窗口中输入文本、修改文本。

✎ 输入命令时要注意看命令行的显示,以免输入错误的命令。若输入错误的命令可用按 Ctrl+G 热键取消。

2.3.2 g++编译器的使用

在 Linux 中,GCC(GNU Compiler Conllection)是一个符合 ANSI 标准的集成化的编译器,通过配置不同的前端模块,能够编译 C、C++、Java 等语言。g++ 是 Linux 中专门编译 C++ 的编译器。

1. g++编译命令选项

g++编译命令有许多选项,按 man g++ 键可以查看编译命令的格式与选项:

[root@localhost yang]# man g++↙

显示命令格式为:

g++[option|filename]

其中 g++ 常用选项如表 2-5 所示。

表 2-5 g++常用选项

选 项 名	作 用
-c	只预处理、编译源程序,不进行连接,生成的是目标文件
-o	指定输出的可执行文件,默认输出文件为 a.out
-g	产生调试信息供 GDB 使用
-Idir	将 dir 目录加到搜索头文件的目录列表中
-Ldir	将 dir 目录加到搜索-l 选项指定的库文件的目录列表中
-llibrary	连接程序在创建可执行文件时包含指定的库文件 library。注意:-l 选项标识

2. 编译一个文件

下列命令编译程序 mycpp.cpp:

[root@localhost yang]# g++ - c mycpp.cpp↙

生成了目标文件 mycpp.o。

可以利用 EMACS 的 shell 命令编译源程序,在输入 "Esc-!" 命令后,在命令行输入:

Shell command: g++ - c mycpp.cpp↙

将完成对 mycpp.cpp 的编译,如果有出错信息,出错信息显示在 EMACS 的窗口中。在修改程序后,使用 Ctrl+X Ctrl+S 热键存盘后再编译。

如果程序没有错误,就可以对生成的目标文件进行连接:

[root@localhost yang]# g++ - o mycppexe mycpp.o↙

生成可执行文件 mycppexe。

实际上,去掉-c 选项,g++可以将编译与连接一起进行。

[root@localhost yang]# g++ - o mycppexe mycpp.cpp↙

3. 连接多个文件

mycpp.cpp 分成了 mycpp_1.cpp 与 mycpp_2.cpp,可以先分别对它们进行编译,然后连接。

[root@localhost yang]# g++ -c mycpp_1.cpp ↙
[root@localhost yang]# g++ -c mycpp_2.cpp ↙

在编译 mycpp_2.cpp 时出现错误,这是因为 Sum() 的函数体没有在 mycpp_2.cpp 中。需要添加函数原型宣告语句:

int Sum(int,int);

源程序经修改编译成功后,将两个程序的目标文件连接生成可执行程序 my2cppexe:

[root@localhost yang]# g++ -o my2cppexe mycpp_1.o mycpp_2.o ↙

同样可以将编译与连接一起进行:

[root@localhost yang]# g++ -o my2cppexe mycpp_1.cpp mycpp_2.cpp ↙

2.3.3　程序的运行

运行可执行文件 mycppexe 的命令为:

[root@localhost yang]# ./mycppexe ↙

2.3.4　查帮助

Linux 系统提供了 man 命令来获得所有 Linux 命令以及 C、C++ 语言的函数帮助信息。"man"为手册"manual"的头字母。

[root@localhost yang]# man␣?␣man ↙

显示 man 命令本身的格式与查询的内容,其中常用的格式为:

```
man 关键字
```

执行此命令,查询出包含该关键字的所有文档。例如:

[root@localhost yang]# man␣?␣sqrt ↙

列出了函数 sqrt 功能的详述。

翻阅 man 文档使用 ↓、↑、PgUp、PgDn 键,使用命令 q 退出。

2.3.5　GDB 调试器的使用

相对于程序的编译错误,运行错误更难以改正。发现程序中运行错误最好的方法是在程序中加入 cout<<"xx";这样的调试语句显示中间结果。GDB 是 GNU Source-Level Debugger 的缩写,是 Linux 下调试 C/C++ 源程序的工具。具有强大的功能。

1. GDB 的启动与退出

输入 gdb 命令即可启动 GDB:

[root@localhost yang]# gdb ↙

GDB 的提示符为(gdb),在提示符后输入 quit 或 q 命令退出。

启动 GDB 时还可以载入调试的程序名,如:

```
[root@localhost yang]# gdb mycppexe
```

☆注意：

☞ GDB 载入的可执行文件在编译时要加调试选项-ggdb：

```
[root@localhost yang]# g++ -ggdb -o mycppexe mycpp.cpp
```

2. GDB 常用命令简介

GDB 的功能非常强大，命令很多，这里仅列出几条常用的基本命令做介绍。

GDB 命令在与其他命令不相混淆的情况下可以只输入前面几个字母，甚至一个字母。

GDB 的求助命令 help 列出命令分类，并可获得某类、某个命令的使用介绍。

```
(gdb) help
```

列出 GDB 的命令分类如下：

aliases、breakpoints、data、files、internals、obscure、running、stack、status、support、tracepoints、user-defines。

```
(gdb) help breakpoint
```

列出 breakpoints 类命令：awatch break catch ignore 等。

```
(gdb) h break
```

列出了 break 命令的功能介绍与使用方法。

GDB 常用基本命令分类如表 2-6 所示。

表 2-6 GDB 常用基本命令

类　　别	命　　令	功　　能
断点命令 breakpoints	watch	设置观察点表达式，当值发生改变时程序停止运行
	break	设置断点，格式为：break 行号
	clear	清除断点，格式为：clear 行号；clear 清除所有断点
文件命令 files	file	载入要调试的文件
	list	列程序，格式为：list 开始行号　结束行号；
数据命令 dada	display	设置表达式，当程序运行到断点时显示表达式的值
	delete display	删除显示表达式的编号
	set	为变量赋值，格式为：set 变量=表达式
程序运行 running	cont	停在断点后继续运行
	jump	从当前位置到指定位置之间取消调试，格式为：jump 行号
	kill	结束当前程序的调试
	next	越过函数执行
	step	陷入函数体内执行

3. GDB 使用实例

启动 GDB：

```
[root@localhost yang]# gdb
```

载入调试文件：

(gdb) file mycppexe ↙

列文件：

(gdb) l 1 ↙　（前一个"l"是"list"的第一个字母，后面的"1"是数字 1）
1　＃include＜iostream＞
2　using namespace std;
3　int Sum(int m,int n=1) {
4　　　int t,s=0;
5　　　if(m>n) { t=m; m=n; n=t;}
6　　　for(int i=m; i<=n; i++)
7　　　　　s=s+i;
8　　　return s;
9　}
10　void main()
(gdb) l ↙
11 {
12　　　int m,n,s;
13　　　cin>>m>>n;
14　　　s=Sum(m,n);
15　　　cout<<"Sum("<<m<<","<<n<<") = "<<s<<endl;
16 }

在第 14 行设置断点：

(gdb) break 14 ↙　（显示断点号为 1）

运行程序：

(gdb) run ↙

输入：2　　7 ↙ 后程序在断点处停下，若输入 next 命令则执行 14 行断点处的函数 Sum()；现输入 step 命令：

(gdb) step ↙ （进入函数 Sum()停在第 4 行：int t,s=0;）
(gdb) next ↙ （执行第 4 行）
(gdb) next ↙ （执行第 5 行：if(m>n){...},停在第 6 行：for(int i=m;i<=n;i++){...}）

此时如果再输入 next 则陷入循环中，输入 cont 命令则程序执行完成。

(gdb) cont ↙ （程序执行完成）

GDB 还可以和 EMACS 一起使用，构成 EMACS＋g+++GDB 的"集成环境"。

第 3 章

C++ 上机实验指导

3.1 上机实验题

3.1.1 实验1 上机环境和 C++ 基础编程练习

1. 实验目的

(1) 了解 Visual C++ 或 CodeBlocks 集成开发环境,或 Linux 下的 C++ 开发环境。
(2) 熟悉集成开发环境的基本编辑命令及其功能键,练习使用常用功能菜单命令。
(3) 学习完整的 C++ 程序开发过程(编译、连接、调试、运行及查看结果)。
(4) 理解简单的 C++ 程序结构。
(5) 掌握 C++ 基本数据类型变量、常量的使用,理解其内存的概念。
(6) 学习 C++ 中各种运算式的使用。

2. 实验内容

编写程序实现求解二次方程的解。
基本要求如下:
(1) 仔细阅读上机实验环境指导,熟悉上机环境。
(2) 设置用户目录路径,编写程序,编译、调试并查看运行结果。
(3) 学习如何根据编译信息定位并排除语法错误和语法警告。
(4) 学习并模仿主教材上的程序编写风格,养成良好编程习惯。

3.1.2 实验2 控制结构编程练习

1. 实验目的

(1) 理解 C++ 程序的控制结构。
(2) 熟练使用条件判断。
(3) 熟练使用各种循环结构。
(4) 进一步提高程序调试与修改编译错误的能力。
(5) 注意提高程序的可读性。

2. 实验内容

编写程序模拟实现以下游戏:
现有两人玩猜拳游戏,每人可用拳头表示 3 种物体(石头(rock)、剪刀(scissors)和布(cloth))中的一种,两人同时出拳,游戏胜负规则如下。
(1) 石头对剪刀:石头赢;
(2) 剪刀对布:剪刀赢;
(3) 布对石头:布赢。

3.1.3　实验 3　函数编程练习

1. 实验目的

(1) 掌握函数的声明、定义方法。

(2) 理解函数参数的传递。

(3) 掌握函数调用方法。

2. 实验内容

(1) 编写一个函数 Take(),该函数返回正整数 n 的第 k 位数字。例如,如果 n 为 543210,则调用函数 Take(n,0)返回数字 0,而调用函数 Take(n,3)返回数字 3。注意,数字的位次顺序为从右到左,从 0 开始。

(2) 编写一个带默认参数的函数,计算 4 次多项式的值,并测试该函数。

3.1.4　实验 4　构造数据类型编程练习

1. 实验目的

(1) 理解数组的概念,掌握数组应用的一般方法。

(2) 理解指针的概念,掌握指针的使用。

(3) 深入理解指针与数组的区别与联系。

2. 实验内容

(1) 编写并测试一个函数,将一个二维数组顺时针旋转 90°。例如将数组

```
1 2 3                7 4 1
4 5 6      转换为     8 5 2
7 8 9                9 6 3
```

(2) 定义一个函数 invert(),将数组 a 中的 n 个整数按相反的顺序存放。程序实现要求如下:

- 用数组作为函数形参实现函数 invert(int A[], int n),函数调用时实参为数组。
- 用数组作为函数形参实现函数 invert(int A[], int n),函数调用时实参为指针。
- 用指针作为函数形参实现函数 invert(int *A, int n),函数调用时实参为数组。
- 用指针作为函数形参实现函数 invert(int *A, int n),函数调用时实参为指针。

(3) 用枚举构造数据类型改写实验 2 中的猜拳游戏。

3.1.5　实验 5　类与对象编程练习

1. 实验目的

(1) 掌握类的定义,根据具体需求设计类。

(2) 深入理解 C++ 中类的封装性。

(3) 会根据类创建各种对象。

(4) 掌握对象的各种成员的使用方法。

(5) 通过定义构造函数实现对象的初始化。

2. 实验内容

定义一个 FDAccount 类,用于描述一个定期存折(fixed deposit),实现现金支取、余额合计、信息显示等。存折基本信息包括账号、账户名称、存款余额、存款期限(以月为单位)、存款利率(以百分点为单位)等。

3.1.6 实验 6　继承与派生编程练习

1. 实验目的

(1) 掌握派生与继承的概念与使用方法。
(2) 运用继承机制对现有的类进行重用。
(3) 掌握继承中的构造函数与析构函数的调用顺序。
(4) 为派生类设计合适的构造函数初始化派生类。
(5) 深入理解继承与组合的区别。

2. 实验内容

设计一个人员类 person 和一个日期类 date，由人员类派生出学生类 student 和教师类 professor，学生类和教师类的数据成员 birthday 为日期类。

3.1.7 实验 7　多态性编程练习

1. 实验目的

(1) 理解多态性的概念。
(2) 掌握如何用虚函数实现动态联编。
(3) 掌握如何利用虚基类。

2. 实验内容

设计一个飞机类 plane，由它派生出歼击机类 fighter 和轰炸机类 bomber，歼击机类 fighter 和轰炸机类 bomber 又共同派生出歼轰机(多用途战斗机)。利用虚函数和虚基类描述飞机类及其派生的类族。

3.1.8 实验 8　类模板编程练习

1. 实验目的

(1) 理解类模板的概念。
(2) 掌握类模板的定义、实例化过程。
(3) 掌握类模板的运用。
(4) 通过类模板进一步理解 C++中代码重用的思想。

2. 实验内容

定义一个通用队列模板类，队列中的数据元素类型可以是字符型、双精度型和其他数据类型，队列的基本操作包括队列的初始化、队列的析构、进队列、出队列、判断队列是否为空和队列是否是满队列，测试该队列。

3.1.9 实验 9　输入/输出流与文件系统编程练习

1. 实验目的

(1) 理解 C++的输入/输出流的概念。
(2) 熟悉 I/O 流的工作过程。
(3) 熟悉各种格式标志和各种格式控制方法。
(4) 分清文本文件和二进制文件的区别。
(5) 掌握二进制文件的输入/输出的步骤与操作。
(6) 会运用文件指针以及各种标志。

2. 实验内容

定义一个学生类,包含学生的学号、姓名和成绩等基本信息,将学生信息写入二进制文件 student.dat 中,实现对学生信息的显示、查询和删除等基本功能。

3.1.10 实验10 string 类字符串处理编程练习

1. 实验目的

(1) 理解 C++ 中的 string 类。

(2) 使用 C++ 标准类库中的 string 定义字符串对象。

(3) 能使用 string 类成员函数、操作符对字符串对象进行各种操作。

2. 实验内容

编写程序,对输入文件中的内容进行分析,统计文件的行数、单词数和每个单词出现的次数。

3.1.11 实验11 异常处理编程练习

1. 实验目的

理解与使用 C++ 的异常处理机制。

2. 实验内容

模拟车站机场的危险品检查机,若未发现危险品,通过;若发现危险品,抛出异常。

3.2 上机实验题参考解答

3.2.1 实验1 上机环境和 C++ 基础编程练习

编写程序实现求解二次方程的解。

【参考解答】

```cpp
#include<iostream>
#include<cmath>
using namespace std;
void main()
{
    float a,b,c;                                    //方程系数
    cout <<"Enter the coefficients of a quadratic equation:"<< endl;
    cout <<"\ta:";
    cin >> a;
    cout <<"\tb:";
    cin >> b;
    cout <<"\tc:";
    cin >> c;
//显示方程
    cout <<"The equation is : "<< a <<" * x * x + "<< b <<" * x + "<< c <<" = 0"<< endl;
    float d = b * b - 4 * a * c;
    float x1 = ( - b + sqrt(d))/(2 * a);            //解一
    float x2 = ( - b - sqrt(d))/(2 * a);            //解二
    cout <<"The solutions are:"<< endl;             //输出解
    cout <<"\tx1 = "<< x1 << endl;
```

```
        cout <<"\tx2 = "<< x2 << endl;
        cout <<"Check:"<< endl;                    //检查解
        cout <<"\ta * x1 * x1 + b * x1 + c = "<< a * x1 * x1 + b * x1 + c << endl;
        cout <<"\ta * x2 * x2 + b * x2 + c = "<< a * x2 * x2 + b * x2 + c << endl;
}
```

运行结果：

```
Enter the coefficients of a quadratic equation:
        a:1↙
        b:3↙
        c:2↙
The equation is : 1 * x * x + 3 * x + 2 = 0
The solutions are:
        x1 = -1
        x2 = -2
Check:
        a * x1 * x1 + b * x1 + c = 0
        a * x2 * x2 + b * x2 + c = 0
```

【思考问题】

① 尝试输入不同的方程系数，会出现什么情况？

② 程序总能够求得方程的解吗？该注意什么？

③ 如果程序出现溢出错误，怎么办？

3.2.2　实验2　控制结构编程练习

现有两人玩猜拳游戏，每人可用拳头表示3种物体（石头、剪刀和布）中的一种，两人同时出拳，游戏胜负规则如下。

(1) 石头对剪刀：石头赢。

(2) 剪刀对布：剪刀赢。

(3) 布对石头：布赢。

【参考解答】

```
#include<iostream>
using namespace std;
#define Player1 0;
#define Player2 1;
#define Tie 2;
void main()
{
    int choice1,choice2,winner;
    cout<<"Choose rock(0),cloth(1),or scissors(2):"<< endl;
    cout<<"Player #1:";           //1号选手选择
    cin>>choice1;
    cout<<"Player #2:";           //2号选手选择
    cin>>choice2;
    switch(choice2-choice1)       //判断胜负结果
```

```cpp
    {
        case 0:
            winner = Tie;
            cout <<"\tYou tied."<< endl;
            break;
        case -1:
        case 2:
            winner = Player1;
            cout <<"\tPlayer #1 wins."<< endl;
            break;
        case -2:
        case 1:
            winner = Player2;
            cout <<"\tPlayer #2 wins."<< endl;
    }
}
```

运行结果：

```
Choose rock(0),cloth(1),or scissors(2):
Player #1:2↙
Player #2:0↙
        Player #2 wins.
```

【思考问题】

① 程序可以用 if…else 嵌套实现吗？如果可以，请加以改写。
② 如果游戏规定比赛连续进行 3 次，赢两次或两次以上者最终获胜，如何改写程序？

3.2.3 实验3 函数编程练习

(1) 编写一个函数 Take()，该函数返回正整数 n 的第 k 位数字。例如，如果 n 为 543 210，则调用函数 Take(n,0) 返回数字 0，而调用函数 Take(n,3) 返回数字 3。注意，数字的位次顺序为从右到左，从 0 开始。

【参考解答】

```cpp
#include <iostream>
using namespace std;
int Take(long, int);
void main()
{
    int n,k;
    cout <<"Enter a integer:";
    cin >> n;
    do {
        cout <<"location:";
        cin >> k;
        cout <<"Digit number "<< k <<" of "<< n <<" is "<< Take(n,k)<< endl;
    }while(k>0);
```

```cpp
}
int Take(long n,int k)
{
    for(int i = 0;i < k;i++)
        n/ = 10;                    //去掉最右边的数字
    return n % 10;
}
```

运行结果：

```
Enter a integer:2467389 ↙
location:3 ↙
Digit number 3 of 2467389 is 7
location:5 ↙
Digit number 5 of 2467389 is 4
location:0 ↙
Digit number 0 of 2467389 is 9
```

（2）编写一个带默认参数的函数，计算 4 次多项式的值，并测试该函数。

【参考解答】

```cpp
#include <iostream>
using namespace std;
double polynomial(double,double,double = 0,double = 0,double = 0,double = 0);
                                                                        //函数原型
void main()
{
    double x;
    cout <<"Please enter x: ";
    cin >> x;
    cout <<"polynomial(x,1) = "<< polynomial(x,1)<< endl;
    cout <<"polynomial(x,1,2) = "<< polynomial(x,1,2)<< endl;
    cout <<"polynomial(x,1,2,3) = "<< polynomial(x,1,2,3)<< endl;
    cout <<"polynomial(x,1,2,3,4) = "<< polynomial(x,1,2,3,4)<< endl;
    cout <<"polynomial(x,1,2,3,4,5) = "<< polynomial(x,1,2,3,4,5)<< endl;
}
double polynomial(double x,double a0,double a1,double a2,double
    a3,double a4)
{
    return a0 + (a1 + (a2 + (a3 + a4 * x) * x) * x) * x;
}
```

运行结果：

```
Please enter x:2.36 ↙
polynomial(x,1) = 1
polynomial(x,1,2) = 5.72
polynomial(x,1,2,3) = 22.4288
polynomial(x,1,2,3,4) = 75.0058
polynomial(x,1,2,3,4,5) = 230.108
```

【思考问题】
① 在 C++ 中使用带有默认形参值的函数有什么好处？
② 在 C++ 中使用带有默认形参值的函数时应该注意什么问题？
③ 使用带有默认形参值的函数时，如果省掉了某个实参，应该注意什么？
④ 能够将这个程序改为用函数模板实现吗？

3.2.4　实验4　构造数据类型编程练习

(1) 编写并测试一个函数，将一个二维数组顺时针旋转 90°。例如将数组

```
1 2 3            7 4 1
4 5 6   转换为    8 5 2
7 8 9            9 6 3
```

【参考解答】

```cpp
#include <iostream>
using namespace std;
const int SIZE = 3;
typedef int Matrix[SIZE][SIZE];
void print(Matrix);
void rotate(Matrix);
void main()
{
    Matrix m = {{1,2,3},{4,5,6},{7,8,9}};
    cout <<"Before matrix is rotated:"<< endl;
    print(m);                            //输出旋转前的二维数组
    cout <<"After matrix is rotated:"<< endl;
    rotate(m);                           //旋转二维数组
    print(m);                            //输出旋转后的二维数组
}
void print(Matrix A)                     //输出数组元素
{
    int i,j;
    for(i = 0;i < SIZE;i++)
    {
        for(j = 0;j < SIZE;j++)
            cout << A[i][j]<<"\t";
        cout << endl;
    }
}
void rotate(Matrix A)                    //旋转二维数组
{
    int i,j;
    Matrix temp;
    for(i = 0;i < SIZE;i++)
    {
        for(j = 0;j < SIZE;j++)
            temp[i][j] = A[SIZE - j - 1][i];
    }
    for(i = 0;i < SIZE;i++)
```

```
        {
            for(j = 0;j < SIZE;j++)
                A[i][j] = temp[i][j];
        }
    }
```

运行结果:

```
Before matrix is rotated:
1       2       3
4       5       6
7       8       9
After matrix is rotated:
7       4       1
8       5       2
9       6       3
```

【思考问题】

① 数组会溢出吗？为什么 C++ 中的数组可能溢出，其实质是什么？

② 改写上面的程序，使得数组溢出，然后思考 C++ 是如何处理数组溢出的。

(2) 定义一个函数 invert()，将数组 a 中的 n 个整数按相反的顺序存放。程序实现要求如下：

- 用数组作为函数形参实现函数 invert(int A[], int n)，函数调用时实参为数组。
- 用数组作为函数形参实现函数 invert(int A[], int n)，函数调用时实参为指针。
- 用指针作为函数形参实现函数 invert(int *A, int n)，函数调用时实参为数组。
- 用指针作为函数形参实现函数 invert(int *A, int n)，函数调用时实参为指针。

要求 1：用数组作为函数形参实现函数 invert(int A[], int n)，函数调用时实参为数组。

【参考解答】

```
#include <iostream>
using namespace std;
void invert(int A[], int n)                    //将数组元素反序存放
{
    int i,j,temp;
    int m = (n-1)/2;
    for(i = 0;i <= m;i++)
    {
        j = n-1-i;
        temp = A[i];
        A[i] = A[j];
        A[j] = temp;
    }
}
void main()
{
    int i;
    int a[10] = {0,1,2,3,4,5,6,7,8,9};
```

```
    for(i = 0;i < 10;i++)
      cout << a[i]<<"\t";
    invert(a,10);
    printf("The array has been reverted:\n");
    for(i = 0;i < 10;i++)
      cout << a[i]<<"\t";
    cout << endl;
}
```

要求2：用数组作为函数形参实现函数 invert(int A[], int n)，函数调用时实参为指针。

【参考解答】

```
#include<iostream>
using namespace std;
void invert(int  A[], int n)               //将数组元素反序存放
{
    int i,j,temp;
    int m = (n - 1)/2;
    for(i = 0;i <= m;i++)
    {
        j = n - 1 - i;
        temp = A[i];
         A[i] = A[j];
         A[j] = temp;
    }
}
void main()
{
    int i;
    int a[10] = {0,1,2,3,4,5,6,7,8,9};
    int  * p;
    for(i = 0;i < 10;i++)
        cout << a[i]<<"\t";
    p = a;
    invert(p,10);
    printf("The array has been reverted:\n");
    for(p = a;p < a + 10;p++)
        cout << * p <<"\t";
    cout << endl;
}
```

要求3：用指针作为函数形参实现函数 invert(int * A，int n)，函数调用时实参为数组。

【参考解答】

```
#include<iostream>
using namespace std;
void invert(int   * A, int n)              //将数组元素反序存放
{
    int * i, * j,temp, * p;
```

```cpp
    int m = (n - 1)/2;
    i = A;
    j = A + n - 1;
    p = A + m;
    for(;i < p;i++,j--)
    {
        temp = *i;
        *i = *j;
        *j = temp;
    }
}
void main()
{
    int i;
    int a[10] = {0,1,2,3,4,5,6,7,8,9};
    for(i = 0;i < 10;i++)
        cout << a[i]<<"\t";
    invert(a,10);
    printf("The array has been reverted:\n");
    for(i = 0;i < 10;i++)
        cout << a[i]<<"\t";
    cout << endl;
}
```

要求 4：用指针作为函数形参实现函数 invert(int * A，int n)，函数调用时实参为指针。

【参考解答】

```cpp
#include<iostream>
using namespace std;
void invert(int *A, int n)                  //将数组元素反序存放
{
    int *i, *j,temp, *p;
    int m = (n - 1)/2;
    i = A;
    j = A + n - 1;
    p = A + m;
    for(;i < p;i++,j--)
    {
        temp = *i;
        *i = *j;
        *j = temp;
    }
}
void main()
{
    int i, *p;
    int a[10] = {0,1,2,3,4,5,6,7,8,9};
    for(i = 0;i < 10;i++)
        cout << a[i]<<"\t";
    p = a;
```

```
        invert(p,10);
        printf("The array has been reverted:\n");
        for(p = a;p < a + 10;p++)
            cout << * p <<"\t";
        cout << endl;
}
```

以上 4 个程序的运行结果均相同。

运行结果：

0	1	2	3	4	5	6	7	8	9
The array has been reverted:									
9	8	7	6	4	5	3	2	1	0

【思考问题】

① 思考指针与数组的密切关系。

② 指针与数组作为函数参数，其各自有何特点？

（3）用枚举构造数据类型改写实验 2 中的猜拳游戏。

【参考解答】

```
#include <iostream>
using namespace std;
enum Choice{rock,cloth,scissors};
enum Winner{Player1,Player2,Tie};
void main()
{
    int n;
    Choice choice1,choice2;
    Winner winner;
    cout <<"Choose rock(0),cloth(1),or scissors(2):"<< endl;
    cout <<"Player #1:";                    //1 号选手选择
    cin >> n;
    choice1 = Choice(n);
    cout <<"Player #2:";                    //2 号选手选择
    cin >> n;
    choice2 = Choice(n);
    switch(choice2 - choice1)               //判断胜负结果
    {
        case 0:
            winner = Tie;   break;
        case -1:
        case  2:
            winner = Player1; break;
        case -2:
        case  1:
            winner = Player2;
    }
    if(winner == Tie)                       //输出胜负结果
```

```
            cout <<"\tYou tied."<< endl;
    else if(winner == Player1)
            cout <<"\tPlayer #1 wins."<< endl;
    else
            cout <<"\tPlayer #2 wins."<< endl;
}
```

运行结果：

```
Choose rock(0),cloth(1),or scissors(2):
Player #1:2 ↙
Player #2:0 ↙
        Player #2 wins.
```

3.2.5 实验5 类与对象编程练习

定义一个FDAccount类，用于描述一个定期存折（fixed deposit），实现现金支取、余额合计、信息显示等。存折基本信息包括账号、账户名称、存款余额、存款期限（以月为单位）、存款利率（以百分点为单位）等。

【参考解答】

```
#include<iostream>
using namespace std;
class FDAccount
{
public:
    //构造函数
    FDAccount(char * ID,char * depositor,double amount, int period,
    double rate);double fetch(char * ID,char * depositor,double amount);
                                    //支取到期存款
    void update();                  //计算当前账户余额
    void show();                    //显示账户基本信息
protected:
    double interest_rate;           //存款利率
private:
    char    * accounts;             //账号
    char    * name;                 //账户名称
    double balance;                 //存款余额
    int     term;                   //存款期限
};
//构造函数实现
FDAccount::FDAccount(char * ID,char * depositor,double amount,int period,double rate)
{
    name = depositor;
    accounts = ID;
    if((amount<0)||(rate<0))
    {
        cout <<"数据不正确!"<< endl;
```

```cpp
        exit(1);
    }
    balance = amount;
    term = period;
    interest_rate = rate;
}
double FDAccount::fetch(char * ID,char * depositor,double amount)
{
    cout <<"账号   "<<"账户名称   "<<"支取金额"<< endl;
    cout << accounts <<"   "<< name <<"       "<< amount << endl;
    balance = balance - amount;                //计算取款后余额
    return balance;
}
void FDAccount::update()
{
balance = balance + balance * (interest_rate/100.0) * (term/12.0);
                                               //计算当前账户余额
}
void FDAccount::show()
{   cout <<"显示账户基本信息: "<< endl;
    cout <<"账号   "<<"账户名称   "<<"期限   "<<"利率"<< endl;
    cout << accounts <<" "<< name <<" "<< term <<" "<< interest_rate << endl;
    cout <<"目前账户余额为: "<< balance << endl;
}
void main()
{
    FDAccount depositor("0034","王涛",10078,18,1.98);
                                               //声明对象并初始化对象
    depositor.show();
    cout << endl;
    cout <<"存款已到期!\n"<< endl;
    depositor.update();                        //计算到期账户余额
    depositor.show();
    cout << endl;
    cout <<"支取存款:"<< endl;
    depositor.fetch("0034","王涛",5000);       //支取存款
    cout << endl;
    depositor.show();                          //显示支取存款后的账户余额
}
```

运行结果:

```
显示账户基本信息:
账号   账户名称   期限   利率
0034   王涛        18    1.98
目前账户余额为: 10078

存款已到期!

显示账户基本信息:
账号   账户名称   期限   利率
0034   王涛        18    1.98
目前账户余额为: 10377.3
```

```
支取存款:
账号    账户名称    支取金额
0034    王涛        5000

显示账户基本信息:
账号    账户名称    期限    利率
0034    王涛        18      1.98
目前账户余额为: 5377.32
```

【思考问题】
① 类与对象有什么关系?
② 如何实现对象的初始化?
③ 如何实现对对象的私有成员的访问?
④ C++如何实现对象的内存分配与类定义的内存分配?
⑤ 改写上面的程序,分析C++中对象的内存分配。

3.2.6 实验6 继承与派生编程练习

设计一个人员类person和一个日期类date,由人员类派生出学生类student和教师类professor,学生类和教师类的数据成员birthday为日期类。

【参考解答】

```cpp
#include <iostream>
#include <string>
using namespace std;
class date                                    //日期类
{
private:
    int year;
    int month;
    int day;
public:
    date()
    {
        cout<<"Birthday:";
        cin>>year>>month>>day;
    }
    void display()
    {
        cout<<year<<" - "<<month<<" - "<<day;
    }
};
class person                                  //人员类
{
protected:
    char * name;
public:
    person();
```

```cpp
};
person::person()
{
    char namestr[50];
    cout <<"Name:";
    cin >> namestr;
    name = new char[strlen(namestr) + 1];
    strcpy(name, namestr);
}
class student:public person                    //学生类
{
private:
    int ID;
    int score;
    date birthday;
public:
    student()
    {
        cout <<"Student ID:";
        cin >> ID;
        cout <<"Student score:";
        cin >> score;
    }
    void display()
    {
        cout <<"The basic information:"<< endl;
        cout << ID <<"\t"<< name <<"\t"<< score <<"\t";
        birthday.display();
        cout << endl;
    }
};
class professor:public person                  //教师类
{
private:
    int NO;
    char major[10];
    date birthday;
public:
    professor()
    {
        cout <<"Teacher ID:";
        cin >> NO;
        cout <<"schoolteaching major:";
        cin >> major;
    }
    void display()
    {
        cout <<"The basic information: "<< endl;
        cout <<"\t"<< NO <<"\t"<< name <<"\t"<< major <<"\t";
        birthday.display();
        cout << endl;
    }
```

```
};
void main()
{
    student stu;
    stu.display();
    professor prof;
    prof.display();
}
```

运行结果：

```
Name:Li_kangming ↙
Birthday:2002 12 21 ↙
Student ID:202009084 ↙
Student score:92 ↙
The basic information:
202009084       Li_kangming     92      2002-12-21
Name:Ding_yue ↙
Birthday:1972 6 24 ↙
Teacher ID:2346 ↙
teaching major:management ↙
The basic information:
       2346      Ding_yue       management      1972-6-24
```

【思考问题】

① 如何实现对类的成员对象的私有数据成员的访问？

② 基类、派生类和类成员对象，其构造函数和析构函数的调用顺序是如何实现的？

3.2.7　实验7　多态性编程练习

设计一个飞机类 plane，由它派生出歼击机类 fighter 和轰炸机类 bomber，歼击机类 fighter 和轰炸机类 bomber 又共同派生出歼轰机（多用途战斗机）。利用虚函数和虚基类描述飞机类及其派生的类族。

【参考解答】

```cpp
#include <iostream>
using namespace std;
class  plane                              //飞机
{
private:
    double aerofoil;                      //机翼
    double  airframe;                     //机身
    double  empennage;                    //尾翼
    double voyage;                        //航程
    int   passenger;                      //乘员数
public:
    plane(double,double,double,double,int);
    virtual void display();
```

};
void plane::display()
{
    cout <<"\t"<< aerofoil <<"\t"<< airframe <<"\t"<< empennage <<"\t"<< voyage <<"\t"<< passenger;
}
plane::plane(double wing,double frame,double tail,double distance,int num)
{
    aerofoil = wing;
    airframe = frame;
    empennage = tail;
    voyage = distance;
    passenger = num;
}
class fighter:public plane   //歼击机
{
private:
    int missile;   //导弹数
public:
    fighter(double,double,double,double,int,int);
    void fight();
    void display();
};
fighter::fighter(double wing,double frame,double tail,double distance,
        int num,int load_missile):plane(wing,frame,tail,distance,num)
{
    missile = load_missile;
}
void fighter::fight()
{
    cout <<"Fight!"<< endl;
}
void fighter::display()
{
    cout <<"This is a fighter!"<< endl;
    plane::display();
    cout <<"\t"<< missile << endl;
}
class bomber:public plane   //轰炸机
{
private:
    double bomb;   //载弹量
public:
    bomber(double, double,double,double,int,double);
    void atack();
    double getbomb();
    void display();
};
bomber::bomber(double wing,double frame,double tail,double distance,
    int num, double load_bomb):plane(wing,frame,tail,distance,num)

```cpp
{
    bomb = load_bomb;
}
void bomber::atack()
{
    cout <<"Atack!"<< endl;
}
double bomber::getbomb()
{
    return bomb;
}
void bomber::display()
{
    cout <<"This is a bomber!"<< endl;
    plane::display();
    cout <<"\t"<< bomb << endl;
}
class   fighter_bomber:virtual public fighter,virtual public bomber
                                    //歼轰机：多用途战斗机
{
public:
    fighter_bomber(double, double,double,double,int,int,double);
    void display();
};
void fighter_bomber::display()
{
    cout <<"This is a fighter_bomber!"<< endl;
    fighter::display();
    cout <<"\t"<< getbomb()<< endl;
    fight();
    atack();
}

fighter_bomber::fighter_bomber(double wing,double frame,double tail,
    double distance,int num,int load_missile,double load_bomb):fighter
    (wing,frame,tail,distance,num,load_missile),bomber(wing,frame,tail,
    distance,num,load_bomb)
{ }
void main()
{
    fighter f(10.0, 6.0, 2.5,1800,1,8);         //歼击机
    f.display();
    bomber b(30,9,6,12000,12,6000);             //轰炸机
    b.display();
    fighter_bomber fb(20,7,3.2,4000,2,6,2500); //歼轰机
    fb.display();
}
```

运行结果：

```
This is a fighter!
        10      6       2.5     1800    1       8
This is a bomber!
        30      9       6       12000   12      6000
This is a fighter_bomber!
This is a fighter!
        20      7       3.2     4000    2       6
        2500
Fight!
Atack!
```

【思考问题】

① 虚函数有什么优点？

② 虚基类有什么作用？

③ 对于程序中可能出现的二义性，有哪些方法可以解决？

3.2.8 实验 8 类模板编程练习

定义一个通用队列模板类，队列中的数据元素类型可以是字符型、双精度型和其他数据类型，队列的基本操作包括队列的初始化、队列的析构、进队列、出队列、判断队列是否为空和队列是否是满队列，测试该队列。

【参考解答】

```cpp
#include <iostream>
using namespace std;
template <class T>
class Queue
{
public:
    Queue(int s = 100):size(s + 1),front(0),rear(0)   //构造函数
    {
        data = new T[size];
    }
    ~Queue()                                          //析构函数
    {
        delete []data;
    }
    void insert(const T&x)                            //进队列
    {
        data[rear++ % size] = x;
    }
    T remove()                                        //出队列
    {
        return data[front++ % size];
    }
```

```cpp
        int is_empty()const                          //判断队列是否为空
        {
            return front == rear;
        }
        int is_full()const                           //判断队列是否为满队列
        {
            return((rear + 1) % size == front);
        }
private:
        int front,rear,size;
        T * data;
};
void main()
{
        Queue < char > q(3);                         //队列中的元素类型为 char 型
        q.insert('A');
        q.insert('B');
        q.insert('C');
        if(q.is_full())
            cout <<"Queue is full."<< endl;
        else
            cout <<"Queue is not full."<< endl;
        cout << q.remove()<< endl;
        cout << q.remove()<< endl;
        q.insert('D');
        q.insert('E');
        if(q.is_empty())
            cout <<"Queue is empty."<< endl;
        else
            cout <<"Queue is not empty."<< endl;
            cout << q.remove()<< endl;
            cout << q.remove()<< endl;
            cout << q.remove()<< endl;
        if(q.is_empty())
            cout <<"Queue is empty."<< endl;
        else
            cout <<"Queue is not empty."<< endl;

        Queue < double > p(3);                       //队列中的元素为 double 型
        p.insert(1.11);
        p.insert(2.22);
        p.insert(3.33);
        cout << p.remove()<< endl;
        cout << p.remove()<< endl;
        p.insert(4.44);
        cout << p.remove()<< endl;
}
```

运行结果：

```
Queue is full.
A
B
Queue is not empty.
C
D
E
Queue is empty.
1.11
2.22
3.33
```

3.2.9 实验9 输入/输出流与文件系统编程练习

定义一个学生类，包含学生的学号、姓名和成绩等基本信息，将学生信息写入二进制文件 student.dat 中，实现对学生信息的显示、查询和删除等基本功能。

【参考解答】

```cpp
#include<iostream>
#include<fstream>
using namespace std;
class Student
{
private:
    long No;                                            //学号
    char *Name;                                         //姓名
    int  Score;                                         //成绩
public:
    Student(long = 0,char * = NULL,int = 0);            //构造函数
    long GetNO();
    char *GetName();
    int  GetScore();
    void ShowStudent();                                 //显示学生信息
};
Student::Student(long stu_no,char *stu_name,int stu_score)  //构造函数
{
    No = stu_no;
    Name = stu_name;
    Score = stu_score;
}
void Student::ShowStudent()
{
    cout<<No<<"\t"<<Name<<"\t"<<Score<<endl;
}
long Student::GetNO(){ return No; }
char *Student::GetName(){ return Name; }
int  Student::GetScore(){ return Score; }
```

```cpp
void main()
{
    Student stu[3] = { Student(202007001,"Li ming",70),
                       Student(202007002,"Hu jun",80),
                       Student(202007003,"Wang tian",90)
    };
    int i,k,pos;
    fstream infile,outfile;
    outfile.open("Students.dat",ios::out|ios::binary|ios::trunc);
    if(!outfile)
    {
        cerr<<"File open error!"<< endl;
        exit(1);
    }
    //将学生信息写入 Student.dat 文件
    for(i = 0;i < 3;i++)
    {
        outfile.write((char * )&stu[i],sizeof(stu[i]));
    }
    outfile.close();

    //显示学生信息
    cout <<"Students.dat:"<< endl;
    infile.open("Students.dat",ios::in|ios::binary);
    for(i = 0;i < 3;i++)
    {
        infile.read((char * )&stu[i],sizeof(Student));
        cout << stu[i].GetNO()<<"\t"<< stu[i].GetName()<<"\t"
             << stu[i].GetScore()<< endl;
    }
    infile.close();
    //查询学生信息
    infile.open("Students.dat",ios::in|ios::binary);
    cout <<"Please input the number of record:";
    cin >> k;
    pos = (k - 1) * sizeof(Student);
    infile.seekg(pos);
    infile.read((char * )&stu[i],sizeof(Student));
    cout << stu[i].GetNO()<<"\t"<< stu[i].GetName()<<"\t"
         << stu[i].GetScore()<< endl;
    infile.close();
    //删除第 2 个学生信息
    cout <<"Delete the second record!"<< endl;
    outfile.open("Students.dat",ios::out|ios::binary|ios::trunc);
    for(i = 0;i < 3;i++)
    {
        if(i!= 1)
            outfile.write((char * )&stu[i],sizeof(stu[i]));
    }
    outfile.close();
    //显示删除后的学生信息
    infile.open("Students.dat",ios::in|ios::binary);
```

```
        for(i = 0;i < 2;i++)
        {
            infile.read((char * )&stu[i],sizeof(Student));
            cout << stu[i].GetNO()<<"\t"<< stu[i].GetName()<<"\t"
                << stu[i].GetScore()<< endl;
        }
        infile.close();
    }
```

运行结果：

```
Students.dat:
202007001        Li ming 70
202007002        Hu jun   80
202007003        Wang tian        90
Please input the number of record:2 ↙
202007002        Hu jun   80
Delete the second record!
202007001        Li ming 70
202007003        Wang tian        90
```

3.2.10　实验 10　string 类字符串处理编程练习

编写一个程序，对输入文件中的内容进行分析，统计文件的行数、单词数和每个单词出现的次数。

【参考解答】

```cpp
#include <iostream>
#include <string>
#include <iomanip>
#include <fstream>
using namespace std;
void main()
{
    ifstream in("input.txt");
    string s;
    const int SIZE = 1000;                  //输入文件中单词的最大数
    string word[SIZE];                      //记录读入的单词
    int lines = 0;
    int words = 0;
    int i,n = 0, freq[SIZE] = {0};
    char c;
    while(in >> s)
    {
        if(s.length() == 0)
            continue;
        ++words;
```

```
        in.get(c);
        if(c == '\n')
            ++lines;                             //计算行数
        for(i = 0;i < n;i++)
            if(word[i] == s)    break;
        if(i == n)
            word[n++] = s;                       //把单词加入到列表
        ++freq[i];                               //统计单词出现的频率
    }
    cout <<" The input had "<< lines <<" lines and "<< words <<" words,\nwith the following frequencies:\n";
    for(i = 0;i < n;i++)
    {
        s = word[i];
        if(i > 0 && i % 3 == 0)                  //每行统计 3 个单词
            cout << endl;
        cout << setw(16)<< setiosflags(ios::right)<< s.c_str()<<":"<< setw(2)<< freq[i];
    }
    cout << endl;
```

运行结果:

```
The input had 6 lines and 34 words,
with the following frequencies:
       Auguries:1           of:3    Innocence:1
        William:1        Blake:1           To:1
            see:1            a:4        world:1
             in:4        grain:1         sand:1
            And:2       Heaven:1         wild:1
        flower,:1         Hold:1     infinity:1
            the:1         palm:1         your:1
          hand,:1     eternity:1           an:1
          hour.:1
```

附: input.txt

Auguries of Innocence

William Blake

To see a world in a grain of sand,

And a Heaven in a wild flower,

Hold infinity in the palm of your hand,

And eternity in an hour.

3.2.11 实验 11 异常处理编程练习

实现方法:设计一个 CheckMachine 函数(类),用一个字符串数组存储危险品词语,输入物品用字符串模拟,CheckMachine 对输入的字符串进行危险品词语搜索,若检索到危险品词语,则抛出异常。

【参考解答】

```cpp
#include <string>
#include <iostream>
using namespace std;
string dangerous[] = {"炸药","炸弹","枪支","刀","鞭炮","汽油","硫酸","酒精"};

string CheckMachine(string bag) throw(string){
    for(int i = 0; i < sizeof(dangerous)/sizeof(string); ++i) {
            if(bag.find(dangerous[i])!= string::npos)
                throw (dangerous[i]);
    }
    return "pass";
}
int main(){
    string bag;
    cout <<"输入包内容进行检查,输入quit退出"<< endl;
    getline(cin,bag);                              //接收含空格的字符串
    while(bag!= "quit") {                          //输入quit结束检查
        try {
            cout << CheckMachine(bag)<< endl;
        }
        catch(string s){
            cerr <<"行李中含有:"<< s << endl;
        }
        getline(cin,bag);
    }
}
```

运行结果：

```
输入包内容进行检查,输入quit退出
笔记本电脑1台,衣服2套,书2本,剃须刀1把
行李中含有:刀
烟1条,白酒2瓶,饼干1盒
pass
烟花1盒,鞭炮1盒,火腿1支
行李中含有:鞭炮
quit

Process returned 0 (0x0)   execution time : 366.897 s
Press any key to continue.
```

第4章

C++ 综合应用实验

◇ 引言

C++面向对象编程为软件开发提供了便捷的支持。面向对象的软件开发需要经过面向对象的分析、面向对象的设计、面向对象的编程3个阶段。本章以一个"自助图书借阅系统"的开发为例,详细描述面向对象的软件开发过程。

◇ 学习目标

(1) 掌握简单的软件需求分析方法;

(2) 掌握面向对象的分析与设计方法;

(3) 掌握STL的应用;

(4) 掌握文件的读/写应用。

4.1 系 统 分 析

在系统分析阶段要详细了解业务流程、要解决的问题、使用者的功能需求,并在此基础上确定系统目标。

1. 图书借阅系统的业务流程

图书馆中的图书放在书架上,图书以及读者信息存放在计算机数据库中,读者通过计算机完成借阅、还书过程。图书管理员通过计算机对图书记录与读者记录进行管理。

图书管理员的业务如下。

(1) 管理读者账户:添加读者账户,查阅、维护读者信息。

(2) 管理图书:对图书分类、编条码;图书上架、下架;查询、维护图书信息。

读者的业务如下。

(1) 借书:通过互联网查看图书、预约借阅,到图书馆取书、借书。

(2) 还书:将书还到借书处。

业务流程如下:

(1) 图书管理员将新进的图书分类、编码、输入数据库,将图书分类放置到书架上。

(2) 读者通过网络或图书管理员建立读者账号,获得以PIN标识的借书卡。

(3) 读者到书库浏览图书,将图书选到书车上,凭借书卡号码借书。

(4) 还书时只需将图书投入还书窗口,计算机自动扫描条码将图书记录从读者的账户消除。

(5) 图书管理员将图书从还书窗口上架,供读者借阅。

自助图书借阅系统用于此类中小型图书馆,系统流程如图4-1所示。

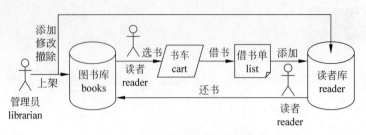

图 4-1 自助图书借阅系统流程图

2. 对象识别

在面向对象的分析中,对象的识别是非常关键的步骤。按面向对象的观点,"世上万事万物皆对象"。在自助图书借阅系统中,诸如书、借书卡、借书单、读者、图书管理员等均为对象。在分析过程中需要寻找、识别出现实世界中有用的对象,并寻找与对象对应的操作(方法)。从自助图书借阅系统的业务流程中识别出相关对象,见表 4-1。

表 4-1 自助图书借阅系统对象

对象	属 性	操作/功能/方法
图书管理员	姓名 [性别,出生年月,住址]	登录,办理借书卡,书籍分类、登记、上架,维护书籍数据,维护读者数据
读者	姓名 借书记录 [性别,出生年月,住址]	查询图书、选书、借书、还书
书	书名,索引号,分类号 作者,出版社,计算机记录	

表中,[ ]标注的是非重要对象、属性、方法。

3. 对象之间的关系与交互

根据业务流程,对象之间的关系与交互设计为:图书管理员建立读者账号、维护图书记录、查询读者记录。读者借阅图书、归还图书。图书管理员与读者及书籍交互,读者与书籍交互。

读者与图书是一对多的关系,即一个读者可以借阅多本图书。

4.2 系统设计

系统设计阶段根据系统分析阶段的结果进行设计,设计出计算机世界中的类以及系统运行流程。

在此阶段,需要添加计算机世界中的类。首先添加 Object 类,作为整个图书借阅系统的基础类;然后添加图书数据类 BookData,描述图书的状态以及状态变迁的操作;添加读者数据类 ReaderData,描述读者的状态以及数据的变化。由于图书数据 BookData 以及读者数据 ReaderData 均存放于数据库中,均需要数据库统一的"增、删、改、查"操作,因此设计一个模板类 Database,提供对数据记录的"增、删、改、查"。此外,设计日期类 Date,提供读当前日期的功能,以便记录借阅日期,判断图书是否超期。

在设计类的过程中,体现"高内聚,低耦合"的原则。内聚就是一个模块内各个元素彼此结合的紧密程度,高内聚就是一个模块内各个元素彼此结合的紧密程度高。耦合是指软件系统结构中各模块之间相互联系紧密程度的一种度量。

系统设计的类图如图 4-2 所示。

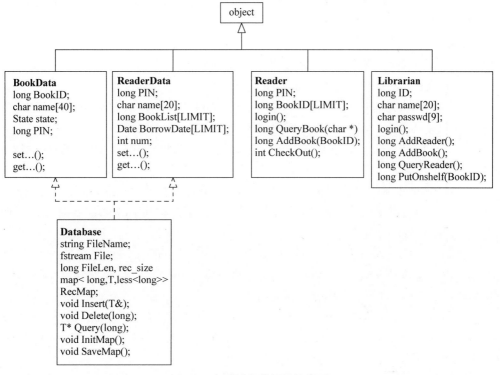

图 4-2　自助图书借阅系统类图

1. BookData 类

BookData 类为图书库的数据类型,BookID 为图书条形码,本系统设计为 long 类型是为了方便操作,在实际应用中应设计成 char [n] 类型。

char name[40] 为书名,设成定长字符便于作为数据库记录。

PIN 为读者借书证号,PIN 字段(属性)与 state 字段共同标识图书在何处。如果 state 的值为 READER,表示图书在号码为 PIN 的读者处。之所以在 BookData 中设置冗余的 PIN 字段,是为了通过图书库快速查找到图书的读者。

State 定义为枚举类型,表示图书状态,取值可以分别为 READER(在读者处)、LIB(在图书馆未上架)、SHELF(在书架上)。

set 与 get 系列方法为标准的设置属性、获得属性的方法。

2. ReaderData 类

ReaderData 类为读者的数据类型,PIN 为读者借书证号,char name[20] 为读者姓名。

BookList[LIMIT] 为读者借阅的书,BorrowDate[LIMIT] 为对应的借阅时间。

num 为借阅的图书数。

3. Database 类

Database 类为数据库模板类,以 BookData 和 ReaderData 为模板参数对其执行数据库操作。

FileName、File 分别为文件名与文件对象。

map<long,T,less<long>> mtype 为 map 容器,关键字为 long 类型,以便存储 BookID、PIN。void Insert(T&)插入记录;void Delete(long)删除记录;T * Query(long)按关键字查询记录;void InitMap()从文件中读数据初始化 map;void SaveMap()将 map 中修改后的数据存到文件中。

4. Reader 类

Reader 类为读者类,当读者登录成功后产生读者对象。

PIN 为登录成功后的借书证号,BookID[LIMIT]存放书车图书。

login()为登录方法,其过程为根据输入(扫描)的借阅者号码从 ReaderData 库中查询读者,若查询到,登录成功(本系统没有要求输入密码,读者自行完成)。

long QueryBook(char *)按输入书名的全部或部分查找出图书,因为在读者找书的过程中图书条码对读者毫无意义。

long AddBook(BookID)将图书放到书车中,此时只需要修改 BookData 库中书的状态。

int CheckOut()借阅书车中的书,此时需要修改 BookData 和 ReaderData 库中的数据。

5. Librarian 类

Librarian 类为图书管理员类,long ID、char name[20]、char passwd[9]分别为管理员号码、姓名、口令。本系统只设了一个管理员口令,在实际应用中有多个管理员,对应多个口令,并且,管理员的户名与口令以文件方式存储。

char login()进行登录;long AddReader()建立读者账户;long AddBook()添加图书;long QueryReader()查询读者信息;long PutOnshelf(BookID)为将图书上架,这些图书可能是读者从书架取下来的,也可能是读者还回来的。

6. 系统运行流程

(1) 建立读者与图书数据库对象,打开读者数据库与图书数据库。

(2) 管理员登录。若登录成功,建立管理员对象。若无图书,则添加图书数据;若无读者数据,建立读者账户;管理员执行其他功能。

(3) 读者登录。若登录成功建立读者对象,选书到书车;借书;读者执行其他功能。

(4) 自动还书。凭输入(扫描)的图书条码修改借书记录。

(5) 系统退出。读者与图书数据库对象消失时将内存中的数据保存到数据库中。

4.3 系统实现

系统实现部分实现各个类以及主函数。

1. 全局数据与 Object 类

```
1    /**************************************************
2     *  程序名: global.h                               *
3     *  功　能: 全局数据与基类                         *
4     **************************************************/
5    const int LIMIT = 3;                        //读者借书限额
6    enum State {READER, LIB, SHELF};
```

```
7   //图书的状态{在读者处、在图书馆(未上架)、在书架上}
8   class object{};
```

2. Date 类的实现

```
1   /**********************************************************
2   *     程序名：date.h                                        *
3   *     功  能：读当前日期、判断日期是否有效                    *
4   **********************************************************/
5   #include<ctime>
6   using namespace std;
7   class Date : public object            //创建一个 Date 类
8   {
9       int year,month,day;
10      int DayOfMonth(int y,int m) const;  //返回一个月的天数
11  public:
12      Date()                             //构造函数,初始化默认的日期
13      {
14          time_t curtime = time(NULL);
15          tm tim = *localtime(&curtime);
16          day = tim.tm_mday;
17          month = tim.tm_mon + 1;
18          year = tim.tm_year + 1900;
19      }
20      Date(int y,int m,int d)
21        :year(y),month(m),day(d)          //对日期赋值,判断有效的日期
22      {
23          if((y<=0)||(m<=0 || m>12)||(d<=0 || d>DayOfMonth(y,m)))
24          {
25              cout<<"Invalid date, data has been set to 1900-1-1"<<endl;
26              year = 1900;
27              month = day = 1;
28          }
29      }
30      virtual ~Date()                    //虚析构函数
31      {
32      }
33      int GetYear()const                 //返回年份
34      {
35          return year;
36      }
37      int GetMonth()const                //返回月份
38      {
39          return month;
40      }
41      int GetDay()const                  //返回日期
42      {
43          return day;
44      }
45      bool IsLeapyear() const            //判断是否为闰年
46      {
```

```cpp
47        return year%400?(year%100?(year%4?false:true):false):true;
48   }
49   bool IsLeapyear(const int y) const          //判断是否为闰年
50   {
51        return y%400?(y%100?(y%4?false:true):false):true;
52   }
53   void display() const                        //输出日期值
54   {
55        std::cout << year <<"-"<< month <<"-"<< day << std::endl;
56   }
57 };
58
59 int Date::DayOfMonth(int y, int m) const
60 {
61      int d = 0;
62      switch(m)
63      {
64         case 1: case 3: case 5: case 7: case 8: case 10: case 12:
65              d = 31;
66              break;
67         case 4: case 6: case 9: case 11:
68              d = 30;
69              break;
70         case 2:
71              d = 28 + IsLeapyear(y);
72              break;
73      }
74      return d;
75 }
```

程序解释：

（1）第 47 行判断 year 是否为闰年，闰年的规则为被 400 整除是闰年，被 4 整除但不被 100 整除是闰年。

（2）第 71 行，如果是闰年，二月多一天（29 天）。IsLeapyear 返回 true，进行算术运算正好是 1。

3. BookData 类的实现

```cpp
1  /**********************************************************
2   *   程序名：BookData.h                                    *
3   *   功  能：图书数据以及 set、get 操作                    *
4   **********************************************************/
5
6  class BookData: public object
7  {
8  private:
9      long BookID;                    //图书条形码
10     char name[40];
11     State state;                    //图书的状态 enum State {READER, LIB, SHELF}
12     long PIN;                       //读者号码
```

```cpp
13      public:
14      BookData(long BookID, const char * name, State state = SHELF, long PIN = 0)
15      {
16          SetID(BookID);
17          SetName(name);
18          SetState(state);
19          SetPIN(PIN);
20      }
21      BookData() {BookID = 0; PIN = 0; }
22      const long GetID() { return BookID;}
23      const char * GetName() { return name;}
24      State GetState() { return state;}
25      long GetPIN() { return PIN;}
26      void ShowData() {cout << BookID <<"\t"<< name <<"\t"
28          << state <<"\t"<< PIN << endl;}
29      void SetID(long BookID) {this －> BookID = BookID;}
30      void SetName(const char * i_name) {strcpy(name, i_name);}
31      void SetState(State state) { this －> state = state;}
32      void SetPIN(long PIN) {this －> PIN = PIN;}
33  };
```

4. ReaderData 类的实现

```cpp
1   /***********************************************************
2    *     程序名: ReaderData.h                                  *
3    *     功  能: 读者数据以及借书、还书操作                      *
4    ***********************************************************/
5   class ReaderData: public object
6   {
7   protected:
8       long PIN;                           //借书证号
9       char name[20];
10      long BookList[LIMIT];               //图书列表
11      Date BorrowDate[LIMIT];             //借书日期
12      int num;                            //借阅图书数
13  public:
14      ReaderData(int PIN, const char * name)
15      {
16          SetID(PIN);
17          SetName(name);
18          num = 0;
19          for(int i = 0; i < LIMIT; ++i)
20              BookList[i] = 0;
21      }
22      ReaderData()
23      {
24          PIN = 0;
25          num = 0;
26          for(int i = 0; i < LIMIT; ++i)
27              BookList[i] = 0;
28      }
```

```cpp
29      void SetID(long PIN){this -> PIN = PIN;}
30      void SetName(const char * i_name) {strcpy(name, i_name);}
31      long GetID() { return PIN;}
32      const char * GetName() { return name;}
33      int GetNum() { return num;}
34      long BorrowBook(long);
35      void ShowData();
36      long ReturnBook(long);
37    };
38  long ReaderData::ReturnBook(long BookID)          //还书
39  {
40      for(int i = 0; i < LIMIT; ++i)
41          if(BookList[i] == BookID)
42          {
43              BookList[i] = 0;
44              -- num;
45              return BookID;
46          }
47      return 0;
48  }
49  void ReaderData::ShowData()
50  {
51      cout << PIN <<"\t"<< name << endl;
52      for(int i = 0; i < LIMIT; ++i)
53      {
54          if(BookList[i])
55          {
56              cout << i + 1 <<": "<< BookList[i]<<"\t";
57              BorrowDate[i].display();
58          }
59      }
60  }
61  long ReaderData::BorrowBook(long BookID)          //借书
62  {
63      for(int i = 0; i < LIMIT; ++i)
64          if(BookList[i] == 0)
65          {
66              BookList[i] = BookID;
67              BorrowDate[i] = Date();
68              num++;
69              return BookID;
70          }
71      cout <<"Book reach the limit!"<< endl;        //图书超过了限额
72      return 0;
73  }
```

5. Database 类的实现

```
1   /**********************************************************
2    *   程序名：Database.h                                    *
3    *   功  能：数据库的打开与保存,数据的插入、查询、删除      *
```

```
4        *************************************************/
5
6    template < class T >
7    class Database: public object
8    {
9      private:
10         fstream File;
11         char FileName[40];                        //文件名
12         long FileLen, rec_size;                   //文件长度,记录大小
13         typedef map < long, T, less < long > > mtype;
14         mtype RecMap;                             //数据记录库
15       public:
16         Database(const char * FileName);
17         ~Database() {SaveMap();}
18         void Insert(T &);
19         void Delete(long);
20         T * Query(long);
21         T * QueryName(const char * );
22         void SaveMap();
23         void ShowAllData();
24    };
25
26    template < class T >
27    Database < T >::Database(const char * FileName)
28    {
29         strcpy(this -> FileName, FileName);
30         File.open(FileName, ios::in | ios::binary);
31         rec_size = sizeof(T);
32         if(File.is_open()) {
33             File.seekg(0, ios::end);
34             if((FileLen = File.tellg()) > 0)
35             {
36                 T Object;
37                 File.seekg(0, ios::beg);
38                 do {
39                     File.read((char * ) & Object, rec_size);
40                     RecMap.insert(mtype::value_type(Object.GetID(), Object));
41                 } while(File.tellg() < FileLen);
42             }
43             File.close();
44         }
45    }
46
47    template < class T >
48    void Database < T >::SaveMap()
49    {
50         mtype::const_iterator iter;
51         T Object;
52         File.open(FileName, ios::out | ios::binary | ios::trunc);
53         for(iter = RecMap.begin(); iter != RecMap.end(); ++iter)
54             File.write((char * ) & iter -> second, rec_size);
55         File.close();
```

```cpp
56  }
57
58  template < class T >
59  void Database < T >::Insert(T & Object)
60  {
61      RecMap.insert(mtype::value_type(Object.GetID(), Object));
62      //cout << typeid(T).name()<<" inserted"<< endl;
63  }
64
65  template < class T >
66  T * Database < T >::Query(long ObjID)
67  {
68      mtype::iterator iter;
69      iter = RecMap.find(ObjID);
70      if(iter == RecMap.end())
71      {
72          cout << ObjID <<" not found!"<< endl;
73          return NULL;
74      } else
75          return &(iter -> second);
76  }
77
78  template < class T >
79  T * Database < T >::QueryName(const char * ObjName)
80  {
81      mtype::iterator iter;
82      for(iter = RecMap.begin(); iter!= RecMap.end();++iter)
83          if(strstr((iter -> second).GetName(),ObjName)!= NULL)
84          {
85              //cout <<"find a name: "<<(iter -> second).GetName()<< endl;
86              return &(iter -> second);
87          }
88      cout << ObjName <<" in "<< typeid(T).name()<<" not found!"<< endl;
89      return NULL;
90  }
91  template < class T >
92  void Database < T >::Delete(long ObjID)
93  {
94      Query(ObjID);
95      RecMap.erase(ObjID);
96  }
97
98  template < class T >
99  void Database < T >::ShowAllData()
100 {
101     mtype::iterator iter;
102     T Object;
103     cout <<"Data in "<< typeid(T).name()<<":"<< endl;
104     for(iter = RecMap.begin(); iter!= RecMap.end(); ++iter) {
105         (iter -> second).ShowData();
106     }
107 }
108
```

程序解释:

(1) 第79行,T * Database<T>::QueryName(const char * ObjName)按输入书名的全部或部分查找出满足条件的第一本书的记录。读者可以完善此方法,查找出满足条件的所有书籍。

(2) 第99行,ShowAllData()显示数据库中的所有记录。在实际应用中,无论是书籍还是读者,库中的记录成千上万,这样的功能用不上。设计此方法用于程序的测试与检验。

6. Librarian 类的实现

```
1   /***********************************************
2    *    程序名:Librarian.h                        *
3    *    功  能:图书管理员登录                     *
4    ***********************************************/
5   class Librarian: public object
6   {
7   private:
8       long ID;                                      //标识
9       char name[20];                                //姓名
10      char passwd[9];                               //口令
11  public:
12      Librarian(long ID, const char * name):ID(ID)
13      {
14          strcpy(this->name,name);
15          strcpy(passwd,"abc");
16      }
17      char login()                                  //登录
18      {
19          char pw[9];
20          for(int i = 0; i < 3; ++i)
21          {
22              cout <<"Enter password:";
23              cin >> pw;
24              if(strcmp(pw,passwd) == 0)
25                  return 'X';
26          }
27          cout <<"Login fail!"<< endl;
28          return 'E';
29      }
30  };
```

程序解释:

程序中只设计了一个口令,在实际应用中有多个管理员,对应多个口令,并且管理员的户名与口令以文件方式存储。

7. Reader 类的实现

```
1   /***********************************************
2    *    程序名:Reader.h                           *
3    *    功  能:读者选书与借书                     *
```

```cpp
4    /******************************************************/
5
6    class Reader: public object
7    {
8    private:
9        long PIN;                                          //读者借书卡号
10       long BookID[LIMIT];                                //书车中存储的书号
11       int num;                                           //书车中书的数目
12   public:
13       Reader(long PIN = 0, const int num = 0):PIN(PIN),num(num) {}
14       int AddBook(const long BookID)                     //将书选入书车中
15       {
16           if(num < LIMIT)
17           {
18               this -> BookID[num] = BookID;
19               cout <<"Book "<< BookID <<" added! "<< endl;
20               num++;
21               return 1;
22           } else
23               cout <<"Cart full!"<< endl;
24           return 0;
25       }
26       long CheckOut() {                                  //借书
27           -- num;
28           return BookID[num];
29       }
30       void ShowCart()
31       {
32           for(int i = 0; i < num; ++i)
33               cout << BookID[i]<< endl;
34       }
35       int GetNum() {return num;}
36   };
```

程序解释:

程序中 CheckOut() 执行的借书功能只是将一本书"移出"了书车，加入到读者数据库中，剩下的操作是由其他模块完成的。

8．main()函数的实现

```cpp
1    /*******************************************
2     *    AutoLibrary.cpp                       *
3     *      自助图书馆                          *
4     *******************************************/
5    # pragma warning(disable:4786)                        //屏蔽4786类型的警告显示
6    # include <iostream>
7    # include <cstring>
8    # include <string>
9    # include <fstream>
10   # include <map>
11   # include "global.h"
```

```cpp
12   #include "date.h"
13   #include "BookData.h"
14   #include "ReaderData.h"
15   #include "database.h"
16   #include "librarian.h"
17   #include "reader.h"
18   using namespace std;
19   int main()
20   {
21       Database<BookData> BookBase("books.dat");          //打开书籍数据库
22       Database<ReaderData> ReaderBase("readers.dat");    //打开读者数据库
23       char choice = 'X';
24       while (!(choice == 'E' || choice == 'e'))
25       {
26           cout<<"(L)ibrarian entry, (R)eader entry, Re(t)urn book, (E)xit:";
27           cin>>choice;
28           Librarian *mgr = NULL;
29           Reader *rdr = NULL;
30           switch(choice)
31           {
32           //==================== 管理员入口 ========================
33           case 'L': case 'l':
34           mgr = new Librarian(101,"yjc");
35           choice = mgr->login(); //成功返回 'X', 否则返回 'E'
36           while (!(choice == 'E' || choice == 'e'))
37           {
38               cout<<"(A)dd reader, Add (B)ooK, (Q)uery Reader,"
39                  <<"(P)ut book to shelf.(E)xit:";
40               cin>>choice;
41               switch(choice)
42               {
43                   long ID;
44                   char name[40];
45                   case 'A': case 'a': //添加读者
46                       cout<<"Give a reader PIN and input a name:";
47                       cin>>ID;
48                       cin.ignore();
49                       cin.get(name,20,'\n');
50                       ReaderBase.Insert(ReaderData(ID,name));
51                       break;
52                   case 'B': case 'b':  //添加书籍
53                       cout<<"Input a book ID and name:";
54                       cin>>ID;
55                       cin.ignore();
56                       cin.get(name,40,'\n');
57                       BookBase.Insert(BookData(ID,name));
58                       break;
59                   case 'Q': case 'q':
60                       cout<<"Input a reader's PIN:";
61                       cin>>ID;
62                       if(ReaderBase.Query(ID) == NULL)
63                           cout<<"No such a reader!"<<endl;
```

```
64              else
65                ReaderBase.Query(ID)->ShowData();
66              break;
67            case 'P': case 'p':
68              cout<<"Input a book ID:";
69              cin>>ID;
70              if(BookBase.Query(ID)==NULL)
71                cout<<"No such a book!"<<endl;
72              else
73                BookBase.Query(ID)->SetState(SHELF);
74              break;
75            case 'S': case 's':
76              ReaderBase.ShowAllData();
77              BookBase.ShowAllData();
78              break;
79            case 'e': case 'E':
80              break;
81            default:
82              cout<<"Unavailable function !\n";
83          }                                                    //结束 switch
84        }                                                      //while
85        delete mgr;
86        choice = 'X';
87        break;
88        //------------------- 管理员退出 ------------------------
89        //================== 读者入口 =========================
90      case 'R': case 'r':
91    long PIN, BookID;
92        int i, t1, t2;
93        char name[40];
94        for(i = 0; i < 3; ++i)
95        {
96          cout<<"Input PIN:";
97          cin>>PIN;
98          if(ReaderBase.Query(PIN)!= NULL)
99          {
100           rdr = new Reader(PIN);
101           break;
102         }
103       }
104       if(i == 3)
105       {
106         cout<<"Check in failed!"<<endl;
107         choice = 'E';
108       }
109       while (!(choice == 'E' || choice == 'e'))
110       {
111         cout<<"(A)dd book to cart, check (O)ut, (Q)uery book by name,"
112           <<"(L)ist my books (E)xit:";
113         cin>>choice;
114         switch(choice)
115         {
```

```cpp
116         case 'A': case 'a':                        //将书添加到书车
117             cout <<"Input a book ID:";
118             cin >> BookID;
119             if((BookBase.Query(BookID)!= NULL)
120                 &&(BookBase.Query(BookID) -> GetState() == SHELF))
121                 if(rdr -> AddBook(BookID))
122                     BookBase.Query(BookID) -> SetState(LIB);
123             break;
124         case 'O': case 'o':                        //退出
125             t1 = rdr -> GetNum();                  /在书车的图书编号
126             t2 = ReaderBase.Query(PIN) -> GetNum();
127             //在读者的图书编号
128             if(t1 > 0 && t2 < LIMIT)
129             {
130                 cout << PIN <<" "<< ReaderBase.Query(PIN) -> GetName()
131                     <<" book list"<< endl;
132                 for(i = 0; i < t1 && i <(LIMIT - t2); ++i)
133                 {
134
135                     BookID = ReaderBase.Query(PIN) -> BorrowBook(rdr -> CheckOut());
136                     BookBase.Query(BookID) -> SetState(READER);
137                     BookBase.Query(BookID) -> SetPIN(PIN);
138                     cout << i + 1 <<"\t"
139                         << BookBase.Query(BookID) -> GetName()<< endl;
140                 }
141                 Date().display();
142             }
143             break;
144         case 'Q': case 'q':
145             cout <<"Input a book name(part):";
146             cin.ignore();
147             cin.get(name,40,'\n');
148             if(BookBase.QueryName(name)!= NULL)
149                 BookBase.QueryName(name) -> ShowData();
150             break;
151         case 'L': case 'l':
152             ReaderBase.Query(PIN) -> ShowData();
153             break;
154         case 'c': case 'C':                        //显示书本
155             rdr -> ShowCart();                     //在书车的图书编号
156             break;
157         case 'e': case 'E':
158             break;
159         default:
160             cout <<"Unavailable function !\n";
161         }                                          //结束 switch
162     }                                              //while
163     delete mgr;
164     choice = 'X';
165     break;
166 //------------------- 读者退出 -----------------------
167 //================== 还书入口 =========================
```

```
168     case 't': case 'T':
169       cout <<"Input a book ID:";
170       cin >> BookID;
171       if(BookBase.Query(BookID)!= NULL)
172       {
173         BookBase.Query(BookID) -> SetState(LIB);
174         if((PIN = BookBase.Query(BookID) -> GetPIN())> 0)
175           ReaderBase.Query(PIN) -> ReturnBook(BookID);
176       }
177       break;
178 //----------------- 退出还书入口 --------------------
179     case 'E': case 'e':
180       break;
181     default:
182       cout <<"Unavailable function !\n";
183     }
184   }
185   return 0;
186 }
```

程序解释：

(1) 程序打开两个数据库后，分成 3 个入口。

- 管理员入口：进入此入口，管理员登录，然后选择添加用户、添加图书、按读者号查询读者信息、按书号查询书籍信息、将书上架等功能。
- 读者入口：进入此入口，读者登录，然后选择按书名查询书籍、选书、借书等功能。
- 还书入口：输入（扫描）书号，将图书还入图书馆。

以上管理员入口以及读者入口的功能在系统设计时分别由 Librarian 类对象和 Reader 类对象完成，而在程序实现时改由在 main()函数中选择不同的入口完成。这是因为读者数据库与书籍数据库是 Librarian 类对象与 Reader 类对象共有，不能各自作为两者的数据成员；如果作为方法中的局部对象，频繁的函数调用需要频繁地将数据库压栈与出栈。

(2) 第 48、55 行用 cin.ignore()过滤掉残留在输入流中的字符。

运行结果：

```
(L)ibrarian entry, (R)eader entry, Re(t)urn book, (E)xit:L↙
Enter password:abc↙
(A)dd reader, Add (B)ooK, (Q)uery Reader,(P)ut book to shelf.(E)xit:a↙
Give a reader PIN and input a name:19771015 Jayden Yang↙
(A)dd reader, Add (B)ooK, (Q)uery Reader,(P)ut book to shelf.(E)xit:a↙
Give a reader PIN and input a name:19731203 Jasmine Yang↙
(A)dd reader, Add (B)ooK, (Q)uery Reader,(P)ut book to shelf.(E)xit:b↙
Input a book ID and name:10001 C programming↙
(A)dd reader, Add (B)ooK, (Q)uery Reader,(P)ut book to shelf.(E)xit:b↙
Input a book ID and name:10002 C++ programming↙
(A)dd reader, Add (B)ooK, (Q)uery Reader,(P)ut book to shelf.(E)xit:b↙
Input a book ID and name:10003 data structure↙
(A)dd reader, Add (B)ooK, (Q)uery Reader,(P)ut book to shelf.(E)xit:b↙
```

```
Input a book ID and name:10004 java programming ↵
(A)dd reader, Add (B)ooK, (Q)uery Reader,(P)ut book to shelf.(E)xit:b ↵
Input a book ID and name:10005 database system concept ↵
(A)dd reader, Add (B)ooK, (Q)uery Reader,(P)ut book to shelf.(E)xit:s ↵
Data in ReaderData:
19731203          Jasmine Yang
19771015          Jayden Yang
Data in BookData:
10001    C programming            2      0
10002    C++ programming          2      0
10003    data structure           2      0
10004    java programming         2      0
10005    database system concept  2      0
(A)dd reader, Add (B)ooK, (Q)uery Reader,(P)ut book to shelf.(E)xit:e ↵
(L)ibrarian entry, (R)eader entry, Re(t)urn book, (E)xit:r ↵
Input PIN:19771015 ↵
(A)dd book to cart, check (O)ut, (Q)uery book by name,(L)ist my books (E)xit:q ↵
Input a book name(part):java ↵
10004    java programming         2      0 ↵
(A)dd book to cart, check (O)ut, (Q)uery book by name,(L)ist my books (E)xit:a ↵
Input a book ID:10004 ↵
Book 10004 added!
(A)dd book to cart, check (O)ut, (Q)uery book by name,(L)ist my books (E)xit:a ↵
Input a book ID:10001 ↵
Book 10001 added!
(A)dd book to cart, check (O)ut, (Q)uery book by name,(L)ist my books (E)xit:a ↵
Input a book ID:10002 ↵
Book 10002 added!
(A)dd book to cart, check (O)ut, (Q)uery book by name,(L)ist my books (E)xit:a ↵
Input a book ID:10003 ↵
Cart full!
(A)dd book to cart, check (O)ut, (Q)uery book by name,(L)ist my books (E)xit:o ↵
19771015 Jayden Yang book list
1        C++ programming
2        C programming
3        java programming
2014-1-1
(A)dd book to cart, check (O)ut, (Q)uery book by name,(L)ist my books (E)xit:e ↵
(L)ibrarian entry, (R)eader entry, Re(t)urn book, (E)xit:t ↵
Input a book ID:10002 ↵
(L)ibrarian entry, (R)eader entry, Re(t)urn book, (E)xit:s ↵
Unavailable function !
(L)ibrarian entry, (R)eader entry, Re(t)urn book, (E)xit:l ↵
Enter password:abc ↵
(A)dd reader, Add (B)ooK, (Q)uery Reader,(P)ut book to shelf.(E)xit:s ↵
Data in ReaderData:
19731203          Jasmine Yang
19771015          Jayden Yang
2: 10001          2014-1-1
3: 10004          2014-1-1
```

```
Data in BookData:
10001    C programming       0          19771015
10002    C++ programming 1              19771015
10003    data structure    2            0
10004    java programming              0          19771015
10005    database system concept 2     0
(A)dd reader, Add (B)ooK, (Q)uery Reader,(P)ut book to shelf.(E)xit:e↙
(L)ibrarian entry, (R)eader entry, Re(t)urn book, (E)xit:e↙
Press any key to continue↙
```

结果解释：

（1）上述程序产生了两个读者，添加了 4 本图书，其中一位读者借了 3 本书，然后还了一本书。

（2）运行结果最后一行要求继续击键退出，如果此时未击键关闭此窗口，main() 函数不能正常退出，Database 的析构函数不能正常执行，将导致数据不能写到磁盘。

本 章 小 结

（1）面向对象软件的开发过程分为面向对象的分析、面向对象的设计、面向对象的编程 3 个过程。

（2）面向对象的分析的主要任务是识别现实世界中的对象，面向对象的设计过程完成类的设计以及类之间的关联以及消息传递设计，这两个过程往往交替进行。

（3）面向对象的设计与编程应体现"高内聚、低耦合"的原则。

（4）一个数据记录较多的软件在实现时应考虑 STL 的使用，并选择合适的容器类。

图书资源支持

感谢您一直以来对清华版图书的支持和爱护。为了配合本书的使用,本书提供配套的资源,有需求的读者请扫描下方的"书圈"微信公众号二维码,在图书专区下载,也可以拨打电话或发送电子邮件咨询。

如果您在使用本书的过程中遇到了什么问题,或者有相关图书出版计划,也请您发邮件告诉我们,以便我们更好地为您服务。

我们的联系方式:

清华大学出版社计算机与信息分社网站:https://www.shuimushuhui.com/

地　　址:北京市海淀区双清路学研大厦 A 座 714

邮　　编:100084

电　　话:010-83470236　010-83470237

客服邮箱:2301891038@qq.com

QQ:2301891038(请写明您的单位和姓名)

资源下载:关注公众号"书圈"下载配套资源。

书圈

清华计算机学堂

观看课程直播